AF522080

TEXTBOOK OF INDUSTRIAL ENGINEERING

By
Shashi Kant Yadav

DISCOVERY PUBLISHING HOUSE PVT. LTD.
NEW DELHI-110 002

Published by:
Tilak Wasan
DISCOVERY PUBLISHING HOUSE PVT. LTD.
4831/24, Ansari Road, Prahlad Street
Darya Ganj, New Delhi-110002 (India)
Phone: +91-11-23279245, 43764432
Fax: +91-11-23253475
E-mail: parul.wasan@gmail.com
info@discoverypublishinggroup.com
discoverypublishinghouse@gmail.com
web: www.discoverypublishinggroup.com

First Edition: **2011**
ISBN: 978-81-8356-841-8

Textbook of Industrial Engineering

Printed at:
Mehra Offset Press
Delhi

Preface

This books "Industrial Engineering" is based on the latest syllabus for B.E./B.Tech. Mechanical and Production Engineering branches of all universities.

This book provides all necessary information about Productivity, Plant Layout, PPC, Maintenance Management, Inventory and Quality Control, Industrial Ownership, Manpower Planning, Organization and job evaluation.

The subject matter has been presented in a simple and systematic manner. To familiarize the student about expected type of questions in the examinations, important summary and review questions have been provided at the end of each chapter.

Suggestions for further improvement of this book are most welcome and would be incorporated in the next edition.

—Author

Contents

1

Analysis of Inventory Problem

INTRODUCTION

A number of factors must be considered in the analysis of inventory problems. Among the most important are the following:

(i) Relevant inventory costs,

(ii) Demand for inventory items,

(iii) Replenishment lead time,

(iv) Length of planning period,

(v) Constraint on the inventory system.

Replenishment Pattern

In a manufacturing system, raw material, semi-finished goods, etc., may be required either to keep them in warehouse for future need or to be used for processing immediately on arrival. The replenishment of such items may be either instantaneous, uniform or constant. These patterns of replenishment depend upon the lead time.

INVENTORY SYSTEM'S PERFORMANCE

The performance of an inventory system is related to the inventory costs. These costs are often not reflected directly in a firm's financial statements. Therefore, system's performance is sometime measured in relation to inventory turnover (annual sales volume divided by the average inventory investment). High inventory turnover indicates a large return on the inventory. But analysis of various costs associated with the inventory of a firm indicates that inventory turnover is not a comprehensive measure of inventory system performance because these costs are affected by the inventory management decisions. Several pertinent inventory costs which are not reflected by the inventory turnover include order preparation cost; cost of carrying inventory, cost of shortage and other customer service costs.

OPERATING CONSTRAINTS

The stock level of various items in the inventory is governed by various constraints such as limited warehouse space, limited budget available for inventory, degree of management attention towards individual items in the inventory, and customer service level (probability of being able to fill a request for product from current stock) to be achieved, etc.

Operating Decision Rules

Two types of managerial decisions need to be made while evolving inventory policies. For a given item, these policies primarily involve the decisions on:

(i) how much (size) is to order (or produce) for each replenishment? and

(ii) when (timing) is it necessary to place an order (or setup production) to replenish inventory?

Decisions regarding the size and timing of replenishment orders are influence'd by four main factors:

The forecast of demand for an item, its replenishment lead time, the inventory related costs for the item and management policies.

Decisions on the size and timing of replenishment of orders for an item can be made by adopting following basic inventory control policies or systems:

CONTINUOUS REVIEW SYSTEMS

- *(s, Q) policy:* Whenever the inventory position (items on hand plus items on order) drops to a given level, s or below, an order is placed for a fixed quantity, Q. This policy is also known as a *fixed-order quantity policy or re-order-point policy*. Here 's' is also termed as reorder point or level; it is also denoted by R.
- *(s, S) policy:* Whenever the inventory position (items on hand plus items on order) drops to a given level, s or below, an order is placed for a sufficient quantity to bring the inventory position upto a given level, S (a pre-determined maximum level).

Periodic Review Systems

- *(T, S) policy:* Inventory position (items on hand plus on order) is reviewed at regular interval of time, spaced at time intervals of length T. At each review, an order is placed for a sufficient quantity to bring the inventory position upto a predetermined maximum level S.

- *(T, s, S) policy:* Inventory position (items on hand plus on order) is reviewed at regular interval of time, spaced at time intervals of length T. At each review, if inventory position is at level, s or below, an order is placed for a sufficient quantity to bring inventory position upto a given level, S; if inventory position is above s, no order is placed. This policy is also known as periodic review policy or fixed-interval policy.

Remarks

1. The (s, S) policy is a special case of the (T, s, S) policy in which T = 0. The (T, s, S) policy can thus be regarded as a periodic version of the (s, S) policy. The (T, S) policy represents a special case of the (T, s, S) policy in which s = S.

   In this chapter we will discuss the determination of order quantity (Q); reorder level (or point) (R) under different conditions and customer service level.

2. The above two policies for replenishment are based on inventory position, rather than simply inventory on hand, on account for cases where the lead time is longer than the time between replenishments.

   If the lead time is always shorter than the time between replenishments, then there will never be any items on order at the time an order is placed; in that case, the review of inventory can be based simply on the inventory on hand.

3. If demand occurs one item at a time, then (s, S) policy is the same as the (s, Q) policy. If, however, demand can occur in batches, so that the inventory position can drop from a level above '↔' to a level below '5' instantaneously (*i.e.*, an overshoot can occur), then the (s, Q) and (s, S) policies are different.

In the (s, Q) policy, the order quantity is fixed and the inventory position just after a replenishment order is placed is variable from one replenishment cycle to another. In the (s, S) policy, the inventory position just after a replenishment order is fixed, and the order quantity is variable.

Relevant Inventory Costs

The costs that are affected (*i.e.*, increase or decrease) by the firm's decision to maintain a particular level of inventory are called relevant costs and definitely play an important role in the study of an inventory system. These costs are classified as follows:

Purchase Cost

This cost consists of the actual price paid for the procurement of items. Its unit of measurement is Rs. per unit. The unit price C of an item is independent of the size of the quantity ordered or purchased (or manufactured). The purchase cost is given by

Purchase cost = (Price per unit) × (Demand per unit time)

= C. D.

When price-break or quantity discounts are available for bulk purchase above a specified quantity, the unit price becomes smaller as size of order, Q exceeds a specified quantity level. In such cases the purchase cost become variable and depends on the size of the order. In this case purchase cost is given by

Purchase cost = Price per unit when order size is

Q × Demand per unit time = C(Q).D.

Carrying (or Holding) Cost

The carrying cost is associated with carrying (or holding) inventory. This cost consists of several costs which includes:

(i) Storage cost for providing warehouse space to store the products,

(ii) Inventory handling cost for payment of salaries to employees,

(iii) Insurance cost against possible loss from fire or other form of damage,

(iv) Opportunity cost of the money invested in inventory,

(v) Obsolescence and deterioration costs when a portion of inventory becomes either obsolete or is lost or pilfered,

(vi) Depreciation, etc.

Carrying cost can be determined by two different ways as given below:

(a) Carrying cost = (Cost of carrying one unit of an item in the inventory for a given length of time) × (Average number of units of an item carried in the inventory for a given length of time)

(b) Carrying cost = (Cost to carry one rupee's worth of inventory per time period) × (Rupee value of units carried)

Further, if r is the carrying charge (or holding cost) as a percentage of average rupee value on an annual basis and C is the unit cost of the item in rupees, then the annual carrying cost for the item may be expressed as percentage of average rupee value of inventory held rather than a rupee carrying cost per unit held as: r . C

Ordering (or Set-up) Cost

Ordering cost is incurred each time an order is placed for procuring items from outside suppliers. The cost per order generally includes :

(i) Requisition cost of handling of invoices, stationery, payments, etc.

(ii) Cost of services which includes cost of mailing, telephone calls and other follow up actions,

(iii) Materials handling cost incurred in receiving, inspecting and storing the items included in the order,

(iv) Accounting and auditing, etc.

When an item is produced internally, ordering cost is referred as set-up cost, which includes both paperwork costs and the physical preparation costs.

Ordering (or set-up) cost is independent of the size of the order (or production), rather it varies with the number of orders placed during a given period of time. Thus, if a large number of orders are placed, more money will be required for procuring the items. Ordering cost may be calculated as given below:

Ordering cost = (Cost per order or per set-up) × (Number of orders or set-ups in the inventory planning period)

ROLE OF INVENTORY FUNCTIONS

The most important function of inventory is to improve profitability through manufacturing and marketing support. Since zero inventory manufacturing-distribution system is not viable, each unit of money invested in inventory should be committed to obtain a specific objective. Other important functions of inventory include geographical specialization, decoupling, balancing supply and demand, and safety stock.

Since investment in inventory represents substantial portion of total capital investment in any business, therefore, questions like, why invest funds in inventory? and what benefit can be derived by investing in inventories? are frequently raised.

Lot-Size (or Cycle) Inventory

It is the inventory necessary to meet the average demand during the successive replenishments. The amount of such inventory depends upon the production lot size, economical shipment quantities, storage space limitation, replenishment lead time, price-quantity discount schedules, and inventory carrying cost, etc.

Geographical Specialization

Due to factors of production such as raw materials, water, power, labour, etc., the economical location of an industry is often at distances from areas of demand. The goods produced from different locations are collected at one place (warehouse/plant) to assemble in final product or to offer a single mixed product shipment to customers. This arrangement leads to economic specialization between manufacturing and distribution of business enterprise.

Pipeline (or Transit) Inventory

Since movement cannot be instantaneous, inventory arises due to shipment of inventory items to distribution centres and customers from production centres. Such an inventory is called a process inventory, as it consists of materials actually being worked on or moving between work centres.

Hence, for satisfying demand without delay, it is essential to hold extra stock at various work places to meet the demand while supply is in transit. The amounts of pipeline inventory depend on the time required for shipment and the nature of the demand.

Safety (or Buffer) Inventory

Some amount of stock is created for protection against uncertainties of demand and the time necessary for delivery of goods, known as lead time. Despite the principle of just-in-time as well as expected customer service needs, the demand and lead time both in general are random variables with known probability distributions.

This may cause an unpredictable shortage with a high penalty cost. Thus to prevent losses due to future uncertainty additional stock have to be held in the form of extra inventories, in addition to the regular stock.

The level of such extra inventory is determined by desirable trade-off between protection against demand and supply uncertainties and the level of investment in safety stock. Such extra stock is known as buffer stock as they provide buffer or a safety margin against future uncertainty.

This function concerns short range variation in either demand or replacement, where a great importance is attributed to inventory planning for determining the size of safety stocks. Safety stocks providing protection against two types of uncertainty:

(a) Concerning with sales in excess of forecast during the replenishment period, and

(b) Concerning delays in replenishment.

The inventories committed to safety stocks constitute the maximum area for improved performance. Safety stocks can be developed by variety of techniques and can be adjusted rapidly in case of error or change in policy.

Seasonal Inventory

Seasonal inventory is needed for products whose markets exhibit seasonal patterns of demand and whose production (or supply) is not uniform, *i.e.,* it varies with time; such as fashion items, agriculture products, children's toys, calendars, etc. In these cases the manufacturer faces peak demand where the production facility is unable to meet the demand on a period-by-period basis. Thus, seasonal (or anticipated) inventories are built up in advance or procured during the period of low demand (provided items are not perishable) to be used in peak demand period. If items are to be produced, then to reduce the strain of peak demand periods upon production facilities, the production of such items may be continued during the low demand period. However, the amount of inventories to be either procured or produced to meet unexpected demand should be determined by balancing the holding (or carrying) and shortage (if any) costs of seasonal inventories.

DECOUPLING INVENTORY

If various manufacturing processes (stages) operate successively, then in the case of breakdown of one, or any disturbance at some stage can affect the entire system. This kind of interdependence is not only costly but also disruptive for the entire system. Thus stocking points of inventory are created between adjacent stages so as to achieve a certain degree of independence in operating the stages. Decoupling refers to breaking of operations such that one operation's supply is independent of another's supply. This provides maximum efficiency of operations within a single facility. The decoupling inventories may be classified into four groups:

(a) *Raw Materials and Component Parts :* The bulk of the stores in terms of volume and value is generally raw materials because it is processed to finished products. Thus, the raw materials inventory is used to decouple the producer from the supplier. This means, raw material and components parts inventory could

(i) act as a buffer to take care of delays on the part of supplier(s), and

(ii) guard against seasonal variations in the demand of final product.

(b) *Work-in-Process :* Inventory Since it takes time to convert raw material into finished product, work- in-process inventory is incurred. This may be due to unbalanced loading of machines, holdups, shortages of tools, etc. This inventory takes the form of orders waiting to be transported between machines or of orders waiting to be processed on a particular machine. The level of such inventory can be changed by changing the manufacturing process, lot sizes or production schedules.

(c) *Finished Goods Inventory:* It is the inventory of final products which could be released for sale to the customers. The size of this inventory depends upon the demand, and the ability of the firm to sell its products, firm's ability to meet customer demand and shelf life of the product and storage capacity.

(d) *Spare Parts Inventory :* These are the parts which are used in the production process but do not become part of the product. The size of the inventory depends on the average life of the components.

An Investment Alternative

Inventory is considered as one of the major areas of asset development, and, thus, should provide a certain minimum return on investment. According to the marginal efficiency of capital (MEC) concept, a firm should invest in those alternatives which provide a greater return than the capital cost.

Each organization is concerned with return-on-investment or return on assets employed (profits/assets). In other words, profits or return on capital (Assets) can be given as

$$= \frac{\text{P rofits}}{\text{Sales}} \times \frac{\text{Sales}}{\text{Capital}} \text{ or } \frac{\text{P rofits}}{\text{Sales}} \times \frac{\text{Sales}}{\text{Assets}}$$

Profits over sales is the profit margin, which depends upon many factors including uncertainties of change. On the other hand, sales over capital is capital turnover. Thus, return-on-investment can be improved by increasing turnover or keeping the assets in inventory low.

Balancing of Supply and Demand

This concerns elapsed time between consumption and production, *i.e.*, balancing inventories should exist to reconcile supply with demand. The most important examples of balancing are seasonal production and year round consumption like woollen textile, air coolers, etc. In a balancing capacity, inventories link the economies of production with consumption's variations. Thus, the balancing function of inventory is concerned with the investment in seasonal stocks and full liquidation within the season.

Objectives of Inventory Control

Inventory consists of a large number of items (may be even in thousands in case of a big industry), and thus, systematic management and control of inventory is quite challenging. The main objectives of inventory control are as following:

1. To obtain an optimum balance between the loss due to non-availability of an item and cost of carrying stocks of products, *i.e.,* maintaining an optimum stock of goods at minimum cost.
2. To determine what to order? When to order? How much to order? And how much to carry in stock? So as to obtain an economy in buying, storing, producing, and selling.
3. To keep the overall investment in inventory at the lowest level, consistent with operating requirements.
4. To consider an inventory as risky investment, *i.e.,* investment in inventory may give higher returns for some item and less for others.
5. To minimize shortage, holding, and replacement costs of inventories leading to maximum efficiency in production and distribution.
6. To maintain waste, surplus, scrap, inactive, and obsolete items at the minimum.
7. To supply the raw materials, sub-assemblies, semi-finished goods, finished goods, etc. to their users according to their specifications at right time and at right price.

Shortage (or Stock Out) and Customer Service Cost

The shortage of items occurs when items cannot be supplied on demand. The shortage can be viewed in two different ways:

(i) The supply of items is awaited by the customers, *i.e.,* the items are back ordered and therefore there is no loss in sale. In this case it is very difficult to determine the nature and magnitude of the back ordering cost. The back ordering cost such as cost of informing customers that the item has been back ordered and expected date of delivery can be easily calculated. But such situation may lead to loss of goodwill and hence will not account for loss of future sales.

(ii) Customers are not ready to wait and, therefore, there is loss of sale. In this case shortage cost shall be measured in terms of goodwill loss and lost profit on each unit of the item demanded but were not available.

One of the frequently used measures for inventory shortage costs is the level of customer service achieved on meeting the product demand, *i.e.,* percentage of demand that is met from inventory upon demand.

Shortage cost in a planning period may be calculated as given below:

Shortage cost=(Cost of being short one unit × (Average number of units short in the inventory) in the inventory)

Total Inventory Cost

If unit of an item depends on the quantity purchase, *i.e.,* price discounts are available, then we should formulate an inventory policy which takes into consideration the purchase cost of the items held in stock also. The total inventory cost (TC) is then given by

Total inventory cost = Purchase cost + Ordering cost + Carrying cost + Shortage cost

When price discounts are not offered, the purchase cost. remains constant and is independent of the quantity purchased. The Total Variable Inventory Cost (TVC) is then given by

Total variable inventory cost = Ordering cost + Carrying cost + Shortage cost.

DEMAND FOR INVENTORY ITEMS

The understanding of the nature of demand (*i.e.,* its rate, size and pattern) for the given inventory item is essential to determine an optimal inventory policy for that item. The size of demand is referred to the number of units of the item required in each period (cycle or season). The size is not measured in terms of the number of units sold because demand may remain unfulfilled due to shortage of sufficient stock or delay in delivery.

The size of demand may be either deterministic or probabilistic. In the deterministic case, the demand over a period of time is known with certainty. This can be fixed (static) or can vary (dynamic) from period to period. But in the probabilistic case, the demand over a period of time is not known with certainty but the nature of such demand can be described by a known probability distribution. The known probability distribution may be either stationary or non-stationary from period to period. These terms are equivalent to static and dynamic demand as explained above in this paragraph. The pattern of demand is the manner in which inventory items are required by the customers. The demand for a given period of time may be satisfied instantaneously at the beginning of the period, or uniformly during that period. The effect of instantaneous and uniform demands reflects directly on the total inventory cost.

Order Cycle

The order cycle is the time period between two successive orders. As mentioned before it may be determined in one of the two ways:

- *Continuous Review :* This is also called perpetual inventory record, so that the number of units of an item on hand is always known. In this case an order of fixed size is placed every time the inventory level reaches at a pre-specified level, called order point or reorder level. This decision rule is also referred to as the two-bin system, fixed order size system or Q-system.
- *Periodic Review :* In this case the orders are placed at equal interval of time, but the size of the order may vary depending on the inventory on hand and on order at the time of the review. This decision rule is also referred to as the fixed order interval system or P-system.

Lead Time or Delivery Lag

When an order is placed, it may require some time before delivery of the items ordered is reached. The time delay between placing an order and receipt.of delivery is called delivery lag or lead time. In general, lead time may be deterministic or probabilistic.

Stock Replenishment

Although, an inventory may operate with lead time, the actual replenishment of stock may occur instantaneously or gradual. Instantaneous replenishment is possible when the stock is purchased from outside sources, while gradual replenishment is possible due to a finite production rate within the firm.

Planning Horizon (or Period)

The planning horizon defines the period over which a particular inventory level will be maintained. This period may be finite or infinite depending on the nature of the demand.

INVENTORY MODEL BUILDING

An inventory control problem can be solved by using several methods starting from trial-and-error methods to mathematical and simulation models. Mathematical models help in deriving certain rules which may suggest how to minimize the total (or incremental) inventory cost in case of deterministic demand or how to minimize expected cost in case of probabilistic demand. We will discuss deterministic inventory control models and also will discuss the derivation of economic order quantity (EOQ) for a given inventory situation.

STEPS OF INVENTORY MODEL BUILDING

The steps to build up a suitable inventory model and then to derive rules are as follows:

Step 1 : Collect the data regarding the pattern of demand, the replenishment policy, planning period, relevant inventory costs, etc.

Step 2 : Build up a mathematical model (or some appropriate relationships) using above information which can be representative of the behaviour (or objective) of the inventory system. Models so developed may have an objective function of minimizing the total inventory costs subject to changes in inventory reorder policy and constraints of limited resources (such as floor space for storage, capital investment, etc.). Thus the model would either be an unconstrained optimization model or a mathematical programming model depending upon whether constraints are imposed or not.

Step 3 : Derive an optimal inventory policy (*i.e.*, economic order quantity) by using an appropriate solution procedure so as to balance among inventory costs.

Order Quantity Decisions and Concept of EOQ

Order quantity decisions affect the amount of inventory to be maintained at various stocking points. Large order quantities may reduce the frequency of orders to be placed to procure inventory items and reduce the total ordering cost. But, this decision will increase the cycle stock inventories and cost of carrying inventories.

The determination of order quantities (or batch size for production) raises the question of what order (or lot) size provides the most economical trade-off between relevant inventory costs, *viz.*, ordering, carrying and shortage costs.

The ordering decision is stated in terms of economic order (or lot size) quantity (EOQ). The concept of EOQ was first developed by Ford W. Harris in 1913 for finding the optimum order quantity in order to balance costs of holding excess stock against that of ordering small quantities too frequently. The model that he developed for finding the optimum order quantity has become the basic economic order quantity (EOQ) formula and it serves as the basis for many of the inventory policies currently used in practice.

EOQ is the optimal quantity to order to replenish inventory so as to achieve the optimum total (or variable) inventory cost during the given period of time.

LIST OF SYMBOLS USED

The following symbols for the development of various models discussed in this chapter. The brackets indicate the unit of measurement of each.

C = purchase (or manufacturing) cost of an item (Rs. per unit)

C_0 = ordering (or set-up) cost per order (Rs. per order),

r = cost of carrying one rupee's worth of inventory expressed in terms of percent of rupee value of inventory (per cent per unit time),

C_h = C r = cost of carrying one unit of an item in the inventory for a given length of time (Rs. per item per unit time),

C_s = shortage cost per unit per time (Rs. per unit time),

D = annual requirement (demand) of an item (units per unit time),

Q = order quantity, *i.e.*, number of units ordered per order (units),

ROL = reorder level (or point), *i.e.*, the level of inventory at which an order is placed units),

LT = replenishment lead time (time period),

n = number of orders per time period (orders per unit time),

t = reorder cycle time, *i.e.*, the time interval between successive orders to replenish (time period),

t_p = production period (time period),

r_p = production rate, *i.e.*, the rate at which quantity Q is added to inventory (quantity per unit time),

TC = total inventory cost (Rs.),

TVC = total variable inventory cost (Rs.).

PROBABILISTIC INVENTORY MODELS

We assumed cost, demand and lead time are all known exactly and are constant. However, if the damand or the lead time or both are not known with certainty. The demand is probabilistic or random variable, following either a discrete or continuous probability distribution. In both the situations, there would be single period model and multiperiod models. These are discussed in detail in the following sections.

Single Period Discrete Probabilistic Demand Model

These models deal with the inventory situation of the items requiring one time purchase only, such as news papers, perishable goods, spare parts and seasonal goods. These items should be stocked at the beginning of a given time period. The demand is unknown, but the probability distribution of demand is given. In these cases, purchases are made only once, *i.e.*, no reordering is possible during the period, if more units are required than stocked. Thus, the lead time factor and ordering costs are least importa t in these models. In single period models, the problem can be studied using marginal (or incremental) or pay-off matrix method. The decision procedure consists of a sequence of steps. In such cases, there are two types of costs involved, namely (a) Over-stocking cost and (b) Under-stocking cost representing opportunity losses incurred when the number of units stocked is not exactly equal to the number of units actually demanded.

The objective is to determine the initial inventory level which will optimize the expected return, taking into account factors such as unit cost, carrying or holding cost, selling price, shortage cost, and salvage value.

The models are discussed using the following notations

$D =$ Number of units of the item demanded (a random variable).

$Q =$ The number of units of the item stocked,

$C =$ Unit cost price,

$C_h =$ The unit carrying cost for the entire period,

$C_s =$ The shortage cost,

$P =$ The unit Selling price,

$S =$ The salvage value.

$C_1 =$ Over-stocking cost or over-ordering cost, (*i.e.*, an opportunity loss associated with each unit left unsold), and

$= C + C_h - V$

$C_2 =$ Under-stocking cost or under-ordering cost (*i.e.*, an opportunity loss due to not meeting the demand).

$= P - C + C_h/2 + C_s$

Single period probabilistic demand model can be discrete or continuous. The discrete problem can be solved through Incremental Analysis Method and Pay-off Matrix Method. These are discussed as following:

ANALYSIS METHOD FOR SINGLE PERIOD DISCRETE PROBABILISTIC INVENTORY MODEL

The cost equation may be developed as following:

(a) If only D units are consumed, for any quantity in the stock Q, for the cost associated with Q units in stock for the specified period of time, is either:

$(Q - D)C_1$, when $D \leq Q$,

or $(D - Q)C_2$ when $D > Q$.

(b) Since, the demand D is random variable, its probability distribution is known. If p(D) denotes the probability of demand (D units), such that total probability is 1, *i.e.,*

$$p(0) + p(1) + ... + p(D) + ... \sum_{D=0}^{\infty} p(D) = 1 \quad ...(1)$$

(c) The sum of expected under stocking cost and overstocking cost would be the total expected cost, say F(Q), which is given by

$$F(Q) = C_1 \sum_{D=0}^{Q}(Q-D)p(D) + C_2 \sum_{D=Q+1}^{\infty}(D-Q)p(D) \quad ...(2)$$

(d) If quantity stocked is optimal, Q^*, the total expected cost $F(Q^*)$ will be minimum. Thus, if one unit more or less than the optimal quanity is stocked, the total expected cost will be higher than the optimal.

For minimum of $F(Q^*)$, the condition

$$\Delta F\ (Q^*-1) < 0 < \Delta F(Q^*) \quad ...(3)$$

must be satisfied

The equation (1.22) can be differenced under the summation sign, for

$$D = Q^*+1$$

$$C_1[(Q^* + 1) - D]p(D)$$
$$= C_2[D - (Q^* + 1)\ p(D)].$$

Which is obviously satisfied here. Now taking first difference following is obtained.

$$F(Q^*) = C_1 \sum_{D=0}^{Q^*} [(Q^* + 1) - (D) - (Q^*-D)]p(D)$$

$$+ C_2 \sum_{D=Q^*+1}^{\infty} [D - (Q^* + 1) - (D - Q^*)]p(D)$$

$$= C_1 \sum_{D=0}^{Q^*} p(D) - C_2\left[\sum_{D=Q^*+1}^{\infty} p(D) - \sum_{D=0}^{Q^*}\right]$$

$$= (C_1 + C_2) \sum_{D=0}^{Q^*} p(D) - C_2 \text{ (Because } \sum_{D=0}^{\infty} = 1) \quad ...(4)$$

From condition (3), $\Delta F\ (Q^*) > 0$

$$(C_1 + C_2) \sum_{D=0}^{Q^*} p(D) - C_2 > 0 \quad ...(5)$$

$$\text{or} \quad \sum_{D=0}^{Q^*} p(D) > \frac{C_2}{C + C_1}$$

Thus, the optimum value of stock level Q^* can be obtained by the relationship

$$\sum_{D=0}^{Q^*-1} p(D) < \frac{C_2}{C + C_1} < \sum_{D=0}^{Q^*} p(D). \qquad \ldots (6)$$

For practical application of equation (6), the three step procedure is as follows:

Step-1: A table showing p(D), the probability and the cumulative probability $p(D \leq Q)$ for each reasonable value of D is prepared.

Step-2: The ratio $C_2/(C_1 + C_2)$ known as service level, is computed.

Step-3: The value of Q satisfying the equation (6) is obtained.

PAY-OFF MATRIX METHOD

This method can be explained by considering again the example-1.8. The trader has five options, *i.e.,* five reasonable strategies. He can stock the items from 4 to 12 units. There is no reason to stock more than 12 units as he can never sell more than 12 and there is no possible reason for ordering less than 2. Since there are five alternative courses of action for stocking and five levels of demand, it follows that there are 5 combinations of one strategy and one level of demand. For these 25 combinations, the trader's payoffs can be determined in the form of pay-off matrix. As per the cost information given, the payoffs are determined for the two situations, *i.e.,*

(a) for the demand not more than the stock level, and,0

(b) for the demand more than the stock quantity, *i.e.*

Pay-offs for	$Q \geq D$	$Q < D$
Cost of items	–50Q	–50Q
Sale of item	100D	100Q
Goodwill cost	–	–30(D – Q)
Salvage value	20 (Q – D)	–
Carrying cost	–5 (Q – D) – 5D/2	–5 Q/2
Total pay-off	–35Q + 82.5D	77.5Q – 30D

The pay-off matrix will be five by five. Each element of the matrix can be determined by above total payoffs for demand less than equal to, or greater than the order size. When demand is less than or equal to the order size, following contributions would be obtained to the pay-off.

(a) The trader purchases the items for Rs. 50Q.

(b) He sells D of them for Rs. 100D.

(c) He earns salvage of Rs. 20(Q – D) for items not sold,

(d) He incurs the carrying cost of Rs. (0.10)(50)(Q – D) on unsold items, and

(e) He incurs an average carrying cost of (0.10)(50)(D/2) on the items sold during the period.

The total pay-off, thus, comes out to be as:

(1) –35Q + 82.5D for demand less or equal to order size.

(2) 77.5Q – 30D for the demand more than the order size.

The pay-off matrix is given as in Table 1.

Table : 1 : Pay-off Matrix

		Units Demanded				(D)
		2	**3**	**4**	**5**	**6**
Units stocked or	4	Rs. 190	130	70	10	–50
order size (Q)	6	120	285	225	165	105
	8	50	215	380	320	260
	10	–20	145	330	475	415
	12	–90	75	240	405	570
Probability of demand		0.30	0.20	0.30	0.15	0.05

The expected pay-off can now be determined for the order size. The procedure for calculating the expected values is as following:

For any given strategy, each possible pay-off for that strategy is multiplied by the corresponding probability of the given level of demand and all these products are added up. Thus, for first strategy of order size 4 units, the expected value of pay-off is:

$$(190)(0.30) + (130)(0.20) + (70)(0.30) + (10)(0.15) + (-50)(0.05)$$

$$= \text{Rs. } 144$$

Proceeding similarly, all the expected values can be calculated as following:

Order size Q	:	4	6	8	10	12
Expected value	:	Rs. 144	190.5	233	214	149.15

The objective is to select the strategy that provides the highest pay-off. Hence, the trader should order for 8 units for the highest expected pay-off of Rs. 233.

Out of the above two methods, incremental analysis provides only the optimum level of purchase quantity and does not indicate about the level of expected profit, whereas pay-off matrix method provides both the answers, *i.e.,* optimum purchase quantity as well as the optimum expected return. Further, it is easy to directly convert the pay-off matrix to the opportunity cost matrix as following:

(i) When the pay-offs are profits, any column of the pay-off matrix corresponding to a specific level of demand is taken and the largest pay-off is selected. Each pay-off in the same column is subtracted from the largest pay-off to get the corresponding opportunity costs.

(ii) When payoffs are costs, the smallest pay-off is taken and smallest pay-off is subtracted from each pay-off in the same column to get the opportunity costs.

The opportunity cost matrix can be formed as in Table 2.

Table 2 : Pay-off Matrix for Opportunity Costs

		Units Demanded(D)				
		4	**6**	**8**	**10**	**12**
Units stocked or	4	Rs. 0	155	310	465	620
order size(Q)	6	70	0	155	310	465
	8	140	70	0	155	310
	10	210	140	70	0	155
	12	280	210	140	70	0
Probability of demand		0.30	0.20	0.30	0.15	0.05

(iii) The expected opportunity costs are determined for each alternative with the objective to select the strategy giving minimum expected opportunity cost. The expected opportuniy cost for the first alternative is

$$(0)(0.30)+(155)(0.20)+(310)(0.30)+(465)(0.15)+(620)(0.05)$$

$$= \text{Rs. } 194.75.$$

Similarly, all the expected values can be calulated as following:

Order size Q	:	4	6	8	10	12
Expected value	:	Rs. 194.75	137.25	94.75*	119.75	178.5

The decision would be to select the minimum expected cost, *i.e.*, the trader should store 8 units for the lowest cost of Rs. 94.75.

A relationship between the expected opportunity costs and expected payoffs can be determined as follows:

Expected opportunity cost (EOC) = K (constant) – expected pay-off (EP)

K = sum of the expected value of the largest entries in each column of the pay-off matrix

$= 190 \times 0.30 + 285 \times 0.20 + 380 \times 0.30 + 475 \times 0.15 + 570 \times 0.05$

$= 327.75$

Thus,

EOC = K(327.75) – Expected pay-off for each strategy ...(1)

From the equation (1), it can be observed that the minimum value of EP will simultaneously produce the minimum of EOC. Thus, the two analyses, *i.e.*, the pay-off matrix method and the opportunity cost matrix method give the same result. If the original pay-off matrix is in terms of cost, the relationship (1) will be of the form:

EOC = EP – K,

where EP is in terms of costs.

FIXED ORDER QUANTITY SYSTEM

This is also known as Re-order Level System, Maximum-Minimum System, Perpetual Inventory System, or Q-System. This is best suited for replenishment of stocks of *A* and *B* class items. In this system, the record of the number of units in inventory is continuously maintained. If the lead time is less than the re-order cycle, an order for fixed quantity *Q* (mostly, it is EOQ) is placed when the inventory level falls to a pre-determined re-order level. .

Operation of this system requires fixing of minimum stock (safety stock), re-order point, and standard re-order quantity.

The minimum stock is intended to absorb random and unpredictable distrubances due to variations in consumption rate, delay in delivery, and rejection in incoming consignments. The re-order level is located some where above the minimum level, the exact location of which depends on the lead time and the demand rate of the item.

The system is quite simple and easy to operate. The daily issues of each item are recorded on its stock card and the balance is worked out, and is compared with its re-order level. If the item drops to or below its re-order level, replenishment action is indicated, otherwise the decision is postponed until the next issue. In order to operate the system efficiently, stock cards must be posted regularly for receipts and issues. Some essentials of this system are:

(a) Incoming materials and materials returned from shops must be carried to the appropriate bins and their stock cards must be posted for "Goods Inward Report" and "Material Return Note" for the quantities received physically.

(b) Materials should be issued in correct quanties.

(c) The balance on the stock cards should be tallied periodically with the physical stock.

(d) Entries must be made on correct cards.

(e) Arithmetical errors in addition and subtraction should be avoided.

(f) Errors in copying and posting quantities on stock card should be cross checked.

The basic parameters required to operate re-order level system are:

(a) *Re-order Level* which equals the sum of minimum stock and the lead time consumption.

(b) *Re-order Quantity* (Q^*) which is fixed and normally equals to the *Economic Order Quantity* (EOQ) or modified economic order quanity (MEOQ).

(c) *Maximum Stock under Re-order Level System* which is the sum of minimum stock and the re-order quantity.

(d) *Average Inventory under Re-order Level System* which in general is either the sum of half of minimum stock and maximum stock or merely the sum of minimum stock and half the re-order quantity.

In this case, an operating principle is derived by taking into account the possibility of a stock-out. Few additional variables to be used are:

U = Random variable representing demand during lead time

σ = Standard deviation of demand during lead time

σ_u = Standard deviation of U = square root of variance U (or Var. U)

Var. $U = (\text{Var.}d)(t^*_L) + (\text{Var.}t_L)(d^*)^2$

U^* = Expected lead time demand, *i.e.*, expected consumption during lead time = (average daily demand) × (average lead time)

t_L = Lead time

t^*_L = Average lead time

Var. t_L = s_{tL} = Variance of lead time

d^* = Average daily demand

I = Average inventory level

σ_d = Standard deviation of daily demand

Var.d= σ^2_d = Variance of daily demand

D^* = Expected annual demand

S = Safety level stock or buffer stock

Z = Number of standard deviations needed for a specified confidence level (*i.e.*, constant of proportionality determined from the normal distribution table corresponding to the expected service level).

R^* = Re-order level

Thus, re-order level is given as,

$R^* = U^* + S$

= Average lead time demand + Safety stock . . . (1)

where,

Safety stock, $S = Z\sigma_u$

σ_U = Standard deviation for lead time consumption

$= (t_L\sigma_d)^{1/2}$ . . . (2)

and lead time consumption, $U^* = d^*t_L$

$$\text{Therefore, order size, } Q^* = \left(\frac{2C_0D}{C_h}\right)^{1/2} \quad \text{. . . (3)}$$

Maximum level = $Q^* + S$ . . . (4)

Average inventory level $I = Q^*/2 + S$ . . . (5)

Various levels for fixed order quantity system can be fixed for the following conditions:

1. Variable Demand and Constant Lead Time,
2. Constant Demand and Variabel Lead Time,
3. Variable Demand and Variable Lead Time.

1. Variable Demand and Constant Lead Time

This can be illustrated through the some examples.

2. Control Levels for Constant Demand and Variable Lead Time

In this case, the safety stock is determined as,

S = (maximum lead time-normal lead time) (demand during lead time)

Furthermore, the levels of safety stock depend upon what the extent of stock-out risk (SOR) is acceptable to an organization. Since it is difficult to obtain the accurate estimate for the storage cost, a reasonable service level (SL) should be specified so that the necessary safety stock can be determined to keep the stock-out risk within the specified limits.

The service level is considered to be the probability of not running out of stock on any stock cycle, *i.e.,* percent of order cycle in which all the demand can be supplied from the stock. Thus

Service level (SL) = 10% – stock-out risk (SOR) and

$$\text{Stock-out risk (SOR)} = \frac{\text{Average number of stock} - \text{out per year}}{\text{Number of cycles per year}}$$

3. Control Levels for Variable Demand and Variable Lead Time

This can be illustrated through the some example:

Periodic Review System

This system is also known as Fixed Period System or Replenishment Inventory System or P-System, and involves the following:

(a) Reviewing of stock levels at a fixed interval of time (review period), and

(b) Placing replenishment orders at the end of each review period.

The replenishment quantity is variable and corresponds to the amount of stock which makes the stock ordered plus the stock on hand equal to a target inventory level. Therefore, there are two key variables for system as following:

(a) The review period. T (fixed time between reviews of inventory records) and

(b) The target inventory level, TI, on which orders are based.

The operation of a periodic review system. This constitutes the following:

(a) An order must be placed at the end of each review period,

(b) Each review period is exactly T days long,

(c) At the end of the period, T, an order is placed for quantity sufficient to replenish inventory to the target level TI.

(d) After the re-order level time t_L, the material arrives and goes into inventory, and

(e) The order quantities in each period are different and are calculated as

Q = TI — inventory on hand—previous orders not yet received.

Where

$$TI = \text{Average Review Period Demand} + \text{Average Lead Time Demand} + \text{Safety Stock}$$

and
$$T = \frac{Q^*}{D^*} = \left[\frac{2C_0}{C_h D^*}\right]^{1/2}$$

In this system also, following three situations may be considered:

1. Control Levels for Variable demand and Constant Lead Time.
2. Control Levels for Constant Demand and Variable Lead Time.
3. Control Levels for Variable Demand and Variable Lead Time.

Control Levels for Variable Demand and Constant Lead Time

The average demand during the lead time and review period is $d^*(T + t_L)$. Safety stock to be held during this period is

$$S = Z\sigma_d (T + t_L)^{1/2} \quad \ldots (6)$$

The target inventory level is

$$TI = d^* (T + t_L) + Z\sigma_d (T + t_L)^{1/2} \quad \ldots (7)$$

The under fixed order quantity system, this is the case of constant lead time and variable demand.

Review period is fixed equal to

$$T = \left[\frac{2 \times 100}{0.2 \times 10 \times 400 \times 365}\right]^{1/2} = 0.026 \text{ Yr.} = 9.5 \text{ days}$$

Lead time = 5 days

Average demand during review period T = 9.5 × 400 = 3800 units

Average demand during lead time t_L = 5×400 = 2000 units

Safety stock = $Z\sigma_d(T + t_L)^{1/2}$

= $1.60 \times 10 \times (9.5 + 5)^{1/2}$ = 61 units

Therefore, target inventory level

TI = 3800 + 2000 + 61 = 5861 units.

Control Levels for Constant Demand and Variable Lead Time

Average review period demand = Td^*

Average lead Time demand = $t^*_L d^*$

Safety stock = S = $Zd^*\sigma_{tL}$ . . . (8)

Thus, $TI = Td^* + t^*_L d^* + Zd^*\sigma_{tL}$

$= (T + t^*_L)d^* + Zd^*\sigma_{tL}$ . . . (9)

where σ_{tL} = standard deviation of lead time

This model can be illustrated with the help of the which is a case of *constant demand and variable lead time.*

For this example, review period is fixed as

T = 1/6 years = 60 days.

Then average demand during review period = 60 × 100 = 6000 units.

Average lead time, t^*_L = 15 days.

Therefore, average demand during average lead time

= 100 × 15 = 1500 units

Variance of lead time = 46 $(days)^2$

Safety stock for the service level of 94.5% confidence

$= 1.60 \times 100\ (46)^{1/2}$

= 1086 units

Therefore, the target inventory level

TI = 6000 + 1500 + 1086 = 8586 units.

Control Levels for Variable Demand and Variable Lead Time

Average review period demand = Td^*

Average lead time demand = $t^*_L d^*$

Safety stock $S = Z[T\ (\text{Var } d) + \text{Var } (U)]^{1/2}$ . . . (10)

where (Var d) $= \sigma^Z_d$ = Variance of daily demand

(Var U) $= \sigma^2_U$ = variance of lead time demand

$= (\text{Var } d)\ (t^*_L) + (\text{Var } t_L)\ (d^*)^2$

Thus,

$TI = (T + t^*_L)d^* + Z[(T + t^*_L)\sigma^2_d + d^{*2}\ \sigma^2_{tL}]^{1/2}$ . . . (11)

The review period *T* is usually fixed as

$$T = \left[\frac{2C_0}{C_h D}\right]^{1/2}$$

Control levels for variable demand and variable lead time under fixed order quantity system.

In this case.

Review period T = 1/6 yr. = 60 days

Thus average consumption during average lead time = 60 × 100 = 6000 units

Average consumption during average lead time = 100 × 15 = 1500

Safety stock $S = Z[T + t^*_L)\sigma d^2 + d^{*2}\sigma t^2_L]^{1/2}$

$= 1.60\,[(60 + 15)\,(80)^2 + (100)^2\,(46)]^{1/2}$

= 1548 units

Therefore TI = 6000 + 1500 + 1548

= 9048 units.

Probabilistic Models (stock-out cost allowed)

The problem can be solved in the following steps:

1. The order size, orders per year, order interval and the daily usage rate are determined under the assumption of certainty ignoring lead time.
2. The average lead time, and the lead time needed are determined for a confidence level of 95% or 99%.
3. EOQ is determined for the lead time obtained above by the following formula:

$$Q_i = \left[\frac{2[C_0 + (ESC)_i]D}{C_h}\right]^{1/2} \quad \text{for } i = 1, 2, \ldots\ldots\ldots\ldots n$$

Where $(ESC)_i$ = Expected stock-out cost for the *i*th lead time provided

4. Total cost for the *i*th order size (Q_i) is determined by the following formula:

$$TC_i = \frac{[C_0 + (ESC)_i]D}{Q_i} + \frac{C_h Q_i}{2} + (EHC)_i$$

Where $(EHC)_i$ is the expected holding cost for the *i*th lead time.

In this case, there would be following three situations:

1. Situation of constant demand and variable lead times,
2. Situation of variable demand rates and fixed lead time, and

3. Situation of variable demand and variable lead time.

Situation of Constant Demand and Variable Lead Times

This situation can be illustrated with the help of the some examples:

Situation of Variable Demand Rates and Fixed Lead Time

In the situation of fixed lead time, the risk of stock-out will depend on the actual demand exceeding the anticipated demand. This risk can be reduced by holding the safety stock. Thus, in such a situation, the problem becomes to select a safety stock against the holding cost.

Situation of Variable Demand and Variable Lead Time

When both the lead time and demand rate change considerably from average values, determination of a satisfactory safety stock becomes more complicated. Direct mathematical techniques are generally not applicable to solve such problems. However, such problems can be solved by the technique, often known as *Monte-Carlo Simulation*. Thus the procedure can be for generating simulated experience through the use of random numbers. The random numbers should be used in a consistent order, *i.e.,* figure can be read in any manner desired by rows or columns, diagonally up or down etc.

PRODUCTION INVENTORY MODEL THROUGH TRANSPORTATION TECHNIQUE

The objective in this case is to devise the minimum-cost of production-inventory plan. The problem can be formulated as 'transportation 'model. The equivalence between the elements of the production and the transportation systems can be established as follows:

	Transportation System		Production System
1.	Source i	1.	Production Period i
2.	Destination j	2.	Demand Period j
3.	Supply at Source i	3.	Production Capacity of Period i
4.	Demand at Destination j	4.	Demand per period j
5.	Transportation Cost from source i to destination j	5.	Production and Inventory Cost from period i to j

Invetory Control

Any kind of resource having economic value and is maintained to fulfil the present and future needs of an organization. Fred Hansman, defined inventory as: An idle resource of any kind provided such a resource has economic value. Such resources may be classified into three categories:

(i) Physical resources such as raw materials, semi-finished goods, finished goods, spare parts, lubricants, etc.

(ii) Human resources such as unused labour (manpower), and

(iii) Financial resources such as working capital, etc.

The following are a few examples of the type of inventory held by various organizations. Since the final product (output) of a service organization such as bank, hospital, etc., cannot be stored for use in the future, the concept of inventory control for them is associated with the various forms of productive capacity.

Type of Organization	Type of Inventories Held
• Manufacturer	Raw materials; spare parts; semi-finished goods; finished goods, etc.
• Hospital	Number of beds; stock of drugs; specialized personnel, etc.
• Bank	Cash reserves; tellers, etc.
• Airline company	Seating capacity; spare parts; specialized maintenance crew, etc.

Inventory of resources is held to provide desirable service to customers (users) and to achieve sales turnover target. Investment in large inventories adversely affects the organization's cash flow and working capital as investment in inventory represents substantial portion of total capital investment in any business. It is, therefore, essential to balance the advantage of having inventory of resources and the cost of maintaining it so as to determine an optimal level of inventory of each resource so that total inventory cost is minimum.

Centuries ago, inventories (in the form of quantity of wheat, number of cattles, amount of gold and silver, etc.,) were viewed as measures of the wealth and power of a country or an individual. Such inventories did not require a scientific management. Till recently, the inventories were also considered as a measure of business failure, and thus, larger emphasis started on the liquidity of assets as inventories. However, at present, fast turnover has become the main goal to be pursued for more profits and less risk.

Since the total investment in all the inventories is quite substantial, inventory control is particularly of great importance in manufacturing organization. A poor inventory control leads to higher costs of production, and thus, a sound scientific control of inventories is an essential requirement of a manufacturing organization.

Presently, inventories are viewed as large potential risk rather than as a measure of wealth due to the fast developments and changes in product life. The present concept of inventories, thus, requires the use of scientific techniques in the management of inventories known as inventory control. Inventory control is the technique of maintaining stock-items at desired level, *i.e.,*, the means by which material of the correct quality and quantity is made available as and when it is required with due regard to economy in the following :

(i) Storage costs.

(ii) Ordering costs,

(iii) Set up costs,

(iv) Manufacturing costs,

(v) Purchase price, and

(vi) Working capital.

ANALYSIS OF INVENTORY CONTROL

Inventory control systems appear complicated, however, there are only a few basic questions to answer for an efficient control of inventory and the most important of these are:

When should an Order be Placed to Replenish Inventory

There are three different approaches to check the status of inventory. First, the periodic review system in which orders are placed at regular intervals of time. The quantity ordered varies, depending on the inventory on hand and on order at the time of review. This system is often used in those situations where the level of inventory is reviewed at the end of certain period and units consumed are replenished.

Fixed order quantity system in which level of inventory is monitored regularly and when it drops to a specified level, a replenishment order for a fixed quantity is placed. Variation in demand is allowed by changing the time between orders.

Optional replenishment systems combine a periodic review system with a minimum order quantity restriction. If the calculated order quantity is less than the minimum forcasted value, no order is placed. This avoids the placement of small orders.

How Much should be Ordered in Each Replenishment

Every time an order is placed there are certain costs incurred on account of administration, transportation, inspection, etc. If large frequent orders are placed, the costs of ordering and delivery are kept low, but stock levels and average inventory value are high; If small frequent orders are placed, the costs of ordering and delivery are high, but average stock level is low. Thus, need is to trade-off these two options to minimize the overall total cost. However, order quantity normally depends on:

- Demand pattern
- Price of an item including discounts for larger orders, impending price rise, etc.
- Lead time
- Various inventory costs.

What Items should be Stock

Holding of stock is expensive so controls are needed to ensure that stock level remains as low as possible. This means:

- Stock of existing items is kept at reasonable level.
- Unnecessary items are not added to the inventory.
- All items which are no longer used are removed from the inventory.

Stocks should be controlled using rational policies to balance between holding cost and demand. That is, comparison should be made between costs and benefits of holding an item in stock. Checks should also be made on usage of items already in stock, and if it is cheaper to discontinue stocking of certain items, then they should be removed as quickly as possible.

Merits and Demerits of Inventory

Inventory is an essential part of business life and are built up to provide a cushion between supply and demand. Hence, inventories are meant for taking care of requirements till next arrival, probable delay in delivery, and sudden increase in demand. Further, inventory is maintained for flow of operations in the manufacturing process where inputs and outputs depend on the market situations of uncertainty. Thus, it is not practically (economically and physically) possible to provide each stock item as and when it is needed. Even if, it is physically possible to provide an item at the point of need, it may be prohibitively expensive. Thus, inventories play an extremely important role because of the following reasons :

1. To receive correct quantity of stock at exact time of requirement to ensure continuous and smooth production.
2. To have economy in buying, *i.e.*, to take advantage of price discount in ordering large quantities.
3. To provide satisfactory customer service through supplying most of the requirements from the stock without delay, *i.e.*, to satisfy demand during period of replenishment.
4. To maintain more stable operating or work force level.
5. To carry reserve stocks to avoid stock-outs to protect the firm against an interruption of the scheduled flow of deliveries.
6. To take advantage of transportation economies.
7. To stabilize production, particularly in the case of seasonal products.
8. To plan overall operation strategy through decoupling of successive stages in the chain of acquiring goods, production of items, transporting to wholesellers and retailers and finally serving to customers.
9. To facilitate economic production runs.
10. To facilitate the intermittent production of several products on the same facility.
11. To keep pace with the changing market conditions, *i.e.*, to provide means to hedge against future price and delivery uncertainties.
12. To make effective use of available capital and/or storage space.
13. To achieve favourable return on investment.

However, some disadvantages associated with excessive inventory are:

1. Holding of larger stocks will lead to blocking of more capital and hence, carrying cost will be higher.
2. There is a greater risk of deterioration and spoilage in longer holding of stocks.
3. Excessive stocks generally result into loss due to obsolescence on account of items becoming unusable due to changes in design or changes in requirements.
4. The insurance cost and property taxes would add to the cost of inventory.

All the above cost elements, if added up, amount to huge expenses, which will be incurred in carrying inventories.

CARRYING INVENTORY REASONS

Inventory has been viewed as a necessary evil (non-earning asset) that cannot be eliminated. It is termed evil because maintaining inventory ties up money which could have been used for alternative purposes and it also increases carrying cost. However, it is considered a necessary investment to achieve workable system of production, distribution and marketing of physical goods. Some of the important reasons for carrying inventory are listed below:

Improve Customer Service

An inventory system usually cannot be designed to economically respond to customers request for products or services in an instantaneous manner. Inventories provide a level of product or service availability which when located in the proximity of the customer, can meet a high customer service level. The presence of these inventories may not only maintain sales, but actually increase it to a certain extent.

Reduce Costs

Holding inventories has a cost associated with it, however, it can indirectly reduce operating costs in other activities and may more than offset the carrying cost. Holding inventories may also encourage economies of production by allowing larger, longer and more production runs. Production output can be decoupled from the variation in demand pattern when inventories exist to act as buffers between the two.

Maintenance of Operational Capability

One of the advantages of keeping inventory is to link various manufacturing stages within the firm so that downtime in one stage does not affect the entire manufacturing process. It helps to keep the production going on by acting as a buffer between successive stages of production.

IRREGULAR SUPPLY AND DEMAND

Variability in time of production and transport goods can cause uncertainties that impact on operating costs as well as customer service levels. Inventories can provide service to customers at various locations by maintaining an adequate supply to meet their immediate and seasonal needs.

Quantity Discounts

Inventory of items is carried to take advantage of price-quantity discount because many suppliers offer discounts for large orders. However, such an advantage must keep a balance between storage cost and higher costs due to spoilage, damaged stock, theft, insurance, etc. Furthermore, investment in high inventory due to bulk purchase will. also reduce cash to be used for other purposes.

Avoiding Stockouts (Shortages)

Labour strikes, natural disasters, variations in demand, and delays in supplies are the type of contingencies against which inventories can afford some protection, as well as to avoid the reputation for constantly being out of stock.

Lost goodwill can be an expensive price to pay for not having the right item at the right time and, therefore, inventories help in avoiding shortage at a minimum cost.

Factors Affecting Inventory

The demand and the costs associated with this are the main aspects of inventory. The various factors associated with these two aspects are discussed below:

Number of Items

The number of items held in inventory influence the competing situation for limited floor space or limited money.

Number of Stages of Inventory

When parts are stocked at more than one stage in the sequential production process, they are called multi-stage inventories.

Availability of Items

Sometimes supply position may be badly affected due to various market conditions which, in turn, affects the inventory position in an enterprise.

GOVERNMENT'S/COMPANY'S POLICY

The Government has laid down certain policy norms for some specific items, such as, imported items, explosive items, highly inflammable items, and other essential items. Similarly, a company may also lay down certain policies based on available capital, space, etc. All these factors influence the level of inventories in any organization.

Among the above factors, the major factors influencing the inventory system are demand, cost factors, and lead time. The remaining other factors are usually considered to be completely known for an inventory system.

ECONOMIC PARAMETERS

The economic parameters generally include the following:

(a) *Production Cost or Purchase Price :* The cost of the product is the amount paid to the supplier (purchase price) for the product received or the direct production cost if manufactured. In the situation of fluctuating prices, planning for inventory depends on the average price (generally considered as fixed price). Thus, the cost/price factor is of great significance in cases where price discounts can be obtained or when large production runs result into lower production cost.

(b) *Procurement Costs :* These are the costs incurred in placing purchase order, known as *ordering costs* or related with the initial preparation of a production system, known as *set up costs.* These costs are directly proportional to number of purchase orders placed or number of set ups made, and are generally considered independent of the quantity ordered or produced. Procurement costs include costs of administration (salaries of persons working for purchasing, tendering, paper work, computer costs, postages, etc.), transportation cost, expediting and follow up costs, inspection cost, processing payments, etc. This is represented as the cost per order or per set up.

(c) *Carrying or Holding Costs :* The costs associated with carrying of stocks of items are considered as *holding costs,* or *storage costs* or *possession costs.* These costs vary directly with the number of items involved. Such costs generally range between 15-30 percent of the price of the concerned product and include the following:

 (i) Storage and handling charges,

 (ii) Interest on capital investment,

 (iii) Insurance expenses,

 (iv) Wages of persons working for this,

 (v) Loss due to pilferage, theft, depreciation, spillage, deterioration, and obsolescence,

 (vi) Costs of safety measures, and

 (vii) Warehouse rents.

Considering all the above costs, the storage cost is represented either as per unit of item held per unit of time or as a percentage of average money value of inventory held.

(d) *Shortage or Stock out Costs :* These are the penalty costs incurred due to stock out situation arising for one reason or the other. This can be on selling side (back-order cost or lost sales costs) as well as on manufacturing side (backlog costs or no backlogging costs). Back order costs and backlog costs arise when the unfilled demand can be satisfied at the later date, *i.e.*, customer waits till he gets the supply. Lost sales costs and no backlogging costs arise when the unfilled demand is lost or the customer does not wait for the supply and goes elsewhere. The shortage costs include the costs of production stoppage, overtime/idle time payments, expediting special orders at higher price, idle machine, loss of goodwill, loss of opportunity to sell, loss of profitability, etc.

(e) *Selling Price :* In some cases, the demand may be influenced by the quantity stocked, and, thus, the inventory model will be based on a profit maximization criterion. This includes the revenue from selling the product. The unit selling price may be constant or variable, depending upon the quantity discount.

(f) *Cost of Operating the Information Processing System :* Records of changes in stock levels can be updated either by hand or by computer. In case, when the inventory levels are not recorded daily, the operating cost is incurred to correctly obtain in the physical counts of inventories. These operating costs are usually fixed over a wide quantity (volume range) of the product.

Demand

The demand pattern of an item may be either *deterministic* or *probabilistic.*

(a) *Deterministic Demand :* In this case, the quantities needed over subsequent periods of time are known with certainty and may be expressed over equal periods of time in terms of known constant demands (static) or in terms of variable demands (dynamic).

(b) *Probabilistic Demand :* This occurs when quantities needed over a certain period of time are not known with certainty, but their pattern can be expressed by a known probability distribution. The probability distribution may be either stationary (static) or non-stationary (dynamic) over time.

The demand of an item for a given period of time may be satisfied instantaneously at the beginning of the period or uniformly during that period. The effect of instantaneous and uniform demand gets directly reflected on the total holding cost of an inventory.

Delivery Lag or Lead Time

The time between the requisition for an item and its receipt is known as *delivery lag or lead time.* In general, the lead time may be deterministic or probabilistic having following four components:

(a) Administrative lead time—can be fixed in nature,

(b) Supplier's lead time—can be fixed in nature,

(c) Transportation lead time—cannot be fixed, and

(d) Inspection lead time—cannot be fixed.

Time Horizon

This refers to the planning period over which inventory is to be controlled. The planning period may be finite or infinite. In general, inventory planning is carried out on annual basis. If the time period is long, the time value of cost should be taken into consideration using proper discount factors. On the other hand, it is also possible that the item can be stored only for a limited time period due to perishability or obsolescence. The model should, therefore, minimize the cost over the specific period of time.

Number of Supply Points (Echlons)

An inventory system may have several stocking points, which are organized such that one point acts as a supply source for some other point. For example, the industry supplies the product to warehouse, which supplies to the retailer, who, in turn, supplies to customer. Each supply point is called an echlon. Multi-echlon inventories include products stocked at various levels in the distribution system.

Government's/Company's Policy

The Government has laid down certain policy norms for some specific items, such as, imported items, explosive items, highly inflammable items, and other essential items. Similarly, a company may also lay down certain policies based on available capital, space, etc. All these factors influence the level of inventories in any organization. Among the above factors, the major factors influencing the inventory system are demand, cost factors, and lead time. The remaining other factors are usually considered to be completely known for an inventory system. The number of items held in inventory influence the competing situation for limited floor space or limited money. When parts are stocked at more than one stage in the sequential production process, they are called multi-stage inventories.

Sometimes supply position may be badly affected due to various market conditions which, in turn, affects the inventory position in an enterprise.

CLASSIFICATION OF INVENTORIES

The inventories can be grouped on the following two basis:

(i) Depending upon the service utility of the inventory, and

(ii) Depending upon the use of inventory.

Classification Depending Upon the Service Utility of Inventory

Depending upon the various service utility aspects of inventory, classification can be made based on manufacturing aspects, service aspects, and control aspects. Thus, accordingly, the inventories can be classified as following (Fig 1):

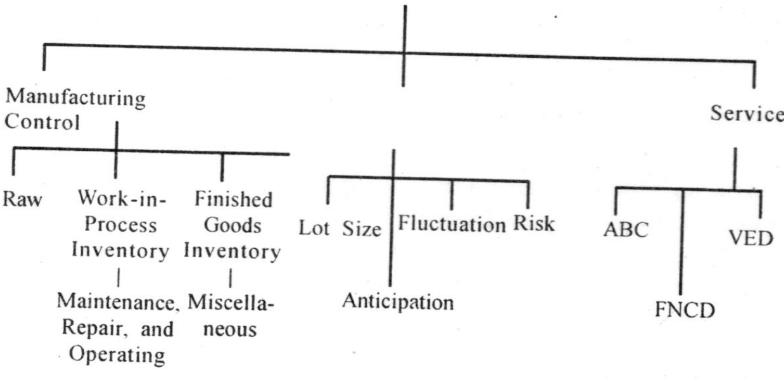

Fig. 1 : Types of Inventory Based on Service Utility.

Manufacturing Inventories

These refer to the inventories held by a manufacturing firm. These can be subdivided as following:

(a) *Production Inventory* : These are the items which go into final product such as raw materials, components and sub-assemblies purchased from outside.

(b) *Work-in-Process Inventory* : This includes all items in semi-finished form or products at different stages of production.

(c) *Finished Goods Inventory* : This includes the products for shipment to users or distributors.

(d) *Maintenance, Repair, and Operating Inventory (M.R.O.)* : These include stores, which do not enter into final product, but are required in the production process.

(e) *Miscellaneous Inventory* : These are the items not covered under above categories. These include obsolete and unsaleable products arising from main production, stationery, and other items required by the office, items required by factory and sales department, cleaning materials, etc.

Service Inventory

These include those inventories, which are required to provide the economic and efficient operations in an organization. These are of four types:

(a) *Lot Size* : This refers to purchases in lot size. This is adopted to (i) obtain quantity discounts, (ii) reduce transportation and purchase costs, and (iii) minimize handling and receiving costs. For example, it would be uneconomical for an oil mill to buy oil seeds every day rather than in bulk during the season of oil seeds.

(b) *Anticipation Stocks* : These are required to meet the predictable changes in demand or in availability of raw materials, *e.g.*, the purchase of wheat in the wheat season for sale of seed preservation products throughout the year.

(c) *Fluctuation Stocks* : These are required to ensure ready supplies to consumers in the event of irregular fluctuations in their demands.

(d) *Risk Stocks* : These are needed to prevent the risk of break-down of production. These are the items, which, usually, need long lead time for supply but are vital and critical for production. This may include some critical spare parts to be imported or needing long time of supply.

INVENTORY PROBLEM CLASSIFICATION

ABC Classification

This concept divides inventories into three categories in terms of percentage of number of items and percentage of total value. This grouping is used in the control of inventory, which is under inventory control techniques. According to this system, inventory items are divided into three categories, *i.e.*, A-consising of few but accounting for major value, B-consising of the medium number of items and involving money also medium, whereas under C, a large number of items but involving less value. This can be represented as shown in Fig. 2.

An inventory control system should be directed towards A-class items because this is the area where major economies can be achieved. Less attention would be required in case of C-class items. ABC analysis helps to concentrate efforts in areas, which need it most and thus, more effective and economical control of inventories is obtained.

A good inventory control system requires a close attention to the F-items and relatively less to N-items. D-items are disposed off.

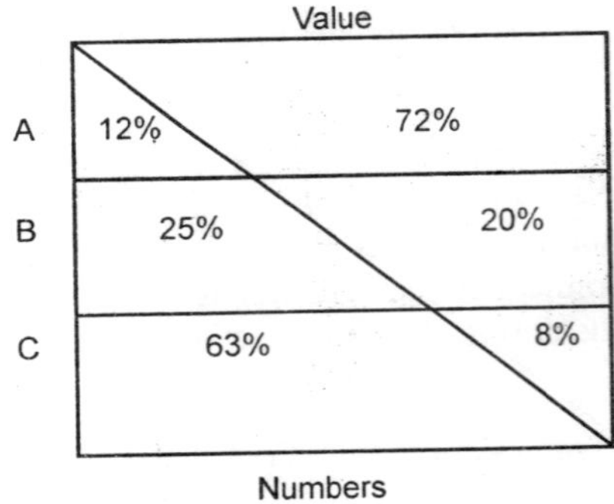

(a) ABC-Classification (Tabular Form)

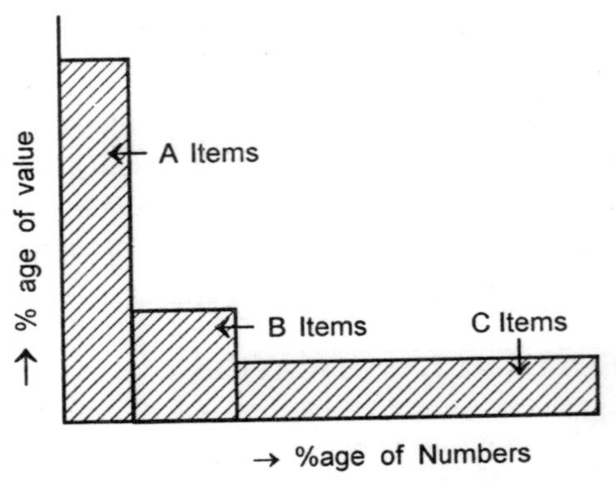

(b) ABC-Classification (Frequency Form)

Fig. 2 : ABC Classification of Inventories.

VED Classification

Sometimes the inventory items for selective control may be classified in the descending order of their criticality as below:

(a) Vital items (or V items),

(b) Essential items (or E items), and

(c) Desirable items (or D items).

If this classification is adopted, a close attention is paid to V items and less to E and least to D items in planning a scheme for inventory control. The criticality of the item may be on technical grounds or on environmental grounds or both. When ABC classification is coupled with VED classification, the efficiency of control on inventories is enhanced. The ABC classification is based on the logic of proportionate value, whereas, VED classification is based on judgement, experience, etc. Under this situation, total number of categories become nine as following:

	V	E	D	Total
A	4	6	1	11
B	7	8	9	24
C	10	35	20	65
Total	21	49	30	100

FNSD Classification

Inventory items may also be classified according to their usage rate as:

(a) Fast moving items (or F items),

(b) Normal moving items (or N items),

(c) Slow moving items (or S items), and

(d) Dead items (or D items).

Classification of Inventory Depending on its Use

Depending on its use, *i.e.,* whether the inventory is used directly in the product, known as *direct inventory,* or the inventory is required in helping the production of the product, known as *indirect inventory,* are discussed below:

Direct Inventories

The items, playing the direct role in the production and becoming an integral part of the finished product, are referred as direct inventories. These may be further classified into following groups:

Raw Materials : These are the basic materials not undergone any conversion but kept in stock for using in the production, since their receipt from the suppliers. Examples are steel (angles, channels, flats, etc.), copper (sheet, tube, etc.), lead, tin, cotton, rubber, timber, etc. These are required for the following purposes:

(a) For economical bulk purchasing,

(b) To enable the changes in production rates,

(c) To serve as buffer stock against delays in transportation, and

(d) To meet the seasonal fluctuations in supplies.

Work in Process Inventories : These are the materials in partially completed condition (semi-finished goods or goods in process) of manufacture. At the end of first operation, raw materials become *work in process* and remain in that classification until they become piece parts or finished goods. Usually the materials on the conveyers, trucks, etc. are considered *work in progress.* These are provided for the following purposes:

(a) To provide economical lot production,

(b) To enable the variety of products,

(c) For replacement of wastages, and

(d) To maintain uniform production even during the variations in sale.

Finished Components Inventories : These include bought out parts or piece parts. Bought out parts are those finished parts or sub-assemblies or assemblies, which are purchased from outside suppliers and generally include standard parts and parts manufactured at supplier's plant to buyer's design. Piece-parts are the parts manufactured at the buyer's own plant from the basic raw materials. These are referred as *semi-finished products.*

Finished Goods Inventories : These are the final products ready for delivery to the customers or finished stock rooms. Products usually leave *"work in progress"* classification and enter *"finished goods"* category at the point of final inspection.

Spare Parts Inventories : These are the parts of main products, which are required for the following:

(a) To provide after sales service to customers, and

(b) To utilize the machine fully by a buyer.

Indirect Inventories

These include the items essentially required for manufacturing, but not becoming an integral part of finished goods. These can be grouped as following:

1. Tools : These consist of

(i) *Standard tools* to be used on machines such as saws, drills, taps, milling cutters, inserts, dies, etc., and

(ii) *Hand tools* such as hand saws, chisels, hammers, needles, spanners, etc.

2. Supplies : These include materials required in running of the plant or in making of company's products, but do not enter into the product. Supplies may include:

(i) *Miscellaneous consumable stores* (such as cotton waste, toilet paper, cleaning powder, etc.)

(ii) *Welding, soldering, and tinning materials* (such as electrodes, gas, welding rods, solder, flux, etc.)

(iii) *Abrasive materials* (such as emery cloth, emery belts, sand paper, etc.)

(iv) *Shipping containers* (such as bags, glass bottles, cardboard boxes, drums, etc.)

(v) *Oils and greases* (such as transformer oil, kerosene oil, petrol, diesel, lubricating oil, etc.)

(vi) *General office supplies* (such as pencils, refills, files, pins, etc.)

(vii) *Electric supplies* (such as cables, fuses, lamps, etc.)

(viii) *Printed forms* (such as envelopes, letter heads, enquiry forms, purchase order forms, etc.)

3. Machinery Spares : These are the materials required in maintenance of machines, *e.g.*, bearings, belts, oil seals, springs, etc.

Inventory Modelling

The methodology for inventory modelling involves the following four steps:

1. The inventory situation is examined carefully, and characteristics and assumptions relating to the situation are listed.
2. Total annual relevant cost equation is developed in narrative form as following:

Total annual relevant cost = +	Cost of the item + costs (purchase cost/ production	Procurement costs (ordering cost/setup cost)	Stock Carrying + stocks + safety stocks) cost)	Stock-out costs (cycle (cost/sales back order)

3. Total annual cost equation is transformed from narrative form into the shorthand logic of mathematics.
4. Finally the cost equation is optimized by finding the optimum for quantity and reorder point and the total relevant cost.

The inventory models may be either deterministic where the variables are known with certainty or stochastic where the variables are probabilistic

SINGLE ITEM INVENTORY MODEL

This refers to the models with all the parameters known with certainty The economic lot size inventory model applies to items which are replenished periodically into inventory in lots covering needs for several periods. The concept of economic lot size is applicable under the following assumptions:

(a) Demand (consumption) of the items is at a constant rate (uniform and continuous) and is known to the decision maker in advance.

(b) The item is replenished in lots or batches, either by purchasing or manufacturing, and

(c) The lead time is also known in advance.

Although, the above assumptions are generally not valid for inventory problems of real situations, simplied models can be developed. into which more realistic and complicating factors can be introduced.

Model-1: The Economic Quantity Model (EOQ) for Instantaneous Supply Case When Stockouts are not Allowed

Under the above assumptions, the model may be described under following situations:

1. Period of demand (planning period) is taken as one year.
2. Demand is known and indicated by D.
3. Purchasing cost or price of one unit is C.
4. Ordering cost (or set up cost) is C_0.
5. Stock holding cost (or inventory carrying cost) is C_h per unit per yean which can be expressed either in terms of cost per period or in terms of percentage charge of the purchase price.
6. Storage cost is C_0 per unit per year.
7. Lead time is L expressed in units of time.
8. Cycle period of replenishment is t.
9. Order size is Q.

When the supply is made immediately after the lead time, the average inventory would be equal to Q/2. Since safety stock is not required for this case, safety stock is zero, *i.e.,* minimum level is zero. This average inventory is affected by the order quantity and the number of orders per year.

In this type of model, there is a set of opposing costs. As the order size increases, the carrying costs will increase, whereas the ordering costs will decrease. On the other hand, as the order size decreases, the carrying costs will decrease, but ordering costs will increase.

Economic order size (EOQ) is that size of order which minimizes total costs of carrying inventory and cost of ordering. Under the conditions of certainty and the annual demand being known, EOQ can be arrived by following three methods:

(a) Tabular Approach,

(b) Graphical Approach, and

(c) Mathematical Approach.

Tabular Approach

This refers to a simple approach for arriving at EOQ by trial and error method. This involves the following three steps:

(a) Selecting a number of possible order sizes,

(b) Determining the total costs for each order size taken, and

(c) Selecting the ordering quantity involving the minimum total cost.

Mathematical Approach

The situation of inventory in this case can be depicted as in Fig. 1.4. Since shortages are not allowed, it implies that shortage cost is prohibitive or C_0 is very large or infinite.

Total annual ordering cost = Total annual carrying cost

or $$C_0 \frac{D}{Q} = C_h \frac{Q}{2} \quad \text{. . . (1)}$$

Thus $$EOQ = Q^* = \left(\frac{2C_0D}{C_h}\right)^{1/2} \quad \text{. . . (2)}$$

i.e., EOQ = Square root of two times of ordering cost multiplied by annual demand divided by carrying cost per unit.

The cycle period is given by $\dfrac{\text{Optimal Order Quantity}}{\text{Annual Demand}}$

or $$t^* = \frac{Q^*}{D} = \left(\frac{2C_0}{C_h D}\right)^{1/2} \quad \ldots (3)$$

The total number of orders per year (=N) is the reciprocal of the cycle,

i.e., $$N = \frac{D^*}{Q} = \left(\frac{C_h}{2C_0}\right)^{1/2} \quad \ldots (4)$$

Total cost = TC + CD + $(2C_0C_hD)$

or Total variable cost (TC*) = $(2C_0C_hD)$ . . . (5)

Lead time consumption = (Lead time in Years)(Demand Rate per Year)

Therefore minimum level = 0, Maximum level = Q*

Re-order Level (R.O.L) = LD.

Model-2: The Economic Quantity Model for Instanteneous Supply Case When Shortages are Allowed

This refers to the situation where shortage cost is finite or it is not large. All the conditions of Model-1 hold good here also. The situation can be shown graphically as shown in Fig. 3.

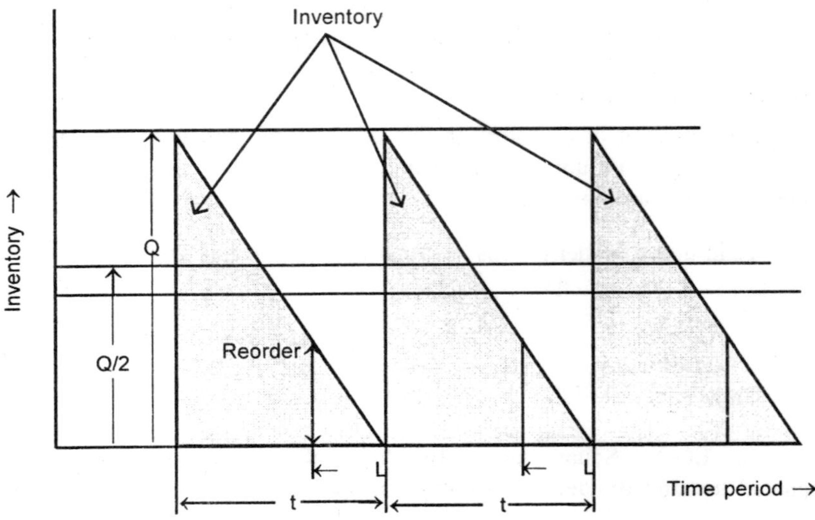

Fig. 3 : Inventory Situation when Shortages are Allowed.

In Fig. 1.4 the triangle ABC represents the inventory and triangle CEF, the shortage. I is the inventory level, and S, the shortage level. Order size Q = I + S. Cycle period $t = t_1 = t_2$ where t_1 is the portion of the cycle period for inventory holding and t_2 is the time of stock-out.

The average number of units in inventory during t_1 is I/2. Therefore, the average inventory cost during $t_1 = (IC_h/2)t_1$.

Where C_h = holding cost per unit for a unit time.

Similarly, the average number of units short during t_2 is

$$\frac{(Q - I)}{2}$$

Therefore, the average cost during

$$t_2 = \frac{(Q - I)}{2} C_s t_2$$

Where C_s = shortage cost per unit for a unit time.

$$\text{Further } t_1 = \frac{I}{Q} t \text{ and } t_2 = \frac{Q - I}{Q} t$$

Let D be the total requirement during a period of time T, and C_0 be the ordering cost per order (or operating cost per production run). Hence, D/Q is the number of orders (production runs) during the period T (say one year). Therefore, the total expected variable cost (TC) during T would be,

$$TC = \left(C_0 + \frac{I}{2} C_h t_1 + \frac{Q-I}{2} C_s t_2\right) \cdot \frac{D}{Q}$$

Substituting the values of t_1 and t_2 obtained earlier, we get

$$TC = \left(C_0 + \frac{I}{2} C_h \frac{I}{Q} t + \frac{Q-I}{2} C_s \frac{Q-I}{Q} . t\right) \frac{D}{Q}$$

$$= \left(C_0 + \frac{I^2}{2Q} C_h t + \frac{(Q-I)^2}{2Q} C_s t\right) \frac{D}{Q} \qquad \ldots (6)$$

Now,

D/Q = number of production runs (or orders) during time T and since t is the time interwal between any two runs (or orders), we have

$$t = \frac{T}{D/Q} = \frac{TQ}{D} \qquad \ldots (7)$$

Substituting the value of t from equation (7) in the expression for TC in equation (6) and simplifying, we get

$$TC = C_0\frac{D}{Q} + \frac{I^2}{2Q}C_hT + \frac{(Q-I)^2}{2Q}C_sT \quad \ldots (8)$$

The optimum values Q and I can be obtained through its partial differentiation of equation (8) with respect to Q and I and equating each equal to zero. Hence,

$$\frac{\delta TC(Q,I)}{\delta I} = \frac{IC_hT}{Q} - \frac{(Q-I)C_sT}{Q} = 0$$

and

$$\frac{\delta TC(Q,I)}{\delta Q} = -\frac{I^2C_hT}{2Q^2} + \left\{\frac{4Q(Q-I) - 2(Q-I)^2}{4Q^2}\right\}C_sT - \frac{C_0D}{Q^2} = 0$$

On simplification of above equations, we get

$$I = Q\frac{C_s}{C_s + C_h}$$

and
$$Q^2C_s - (C_s + C_h)I^2 = \frac{2C_0D}{T}$$

Solving this system of equations for I and Q, we obtain

$$Q = \left[\left(\frac{2DC_0}{TC_h}\right)\left(\frac{C_h + C_s}{C_s}\right)\right]^{1/2}$$

$$I = \left[\left(\frac{2DC_0}{TC_h}\right)\left(\frac{C_s}{C_s + C_h}\right)\right]^{1/2}$$

hence,
$$t = \frac{TQ}{D} = \left[\left(\frac{2TC_0}{DC_h}\right)\left(\frac{C_s + C_h}{C_s}\right)\right]^{1/2}$$

and
$$TC\ \text{Optimum} = \left[(2TDC_hC_0)\left(\frac{C_s}{C_h + C_s}\right)\right]^{1/2}$$

Economic Order Quantity with Price Discounts

Buying in large quantities generally attracts some discounts in price. As a result, the unit price P (Q) of the commodity does not remain constant, rather depends upon the quantity purchased. Buying in large quantities offers the following advantages :

(a) Lower unit cost,

(b) Lower ordering costs,

(c) Fewer stockouts, and

(d) Lower transportation costs.

The disadvantages of buying large quantities are

(a) Higher inventory carrying costs,

(b) More capital investment, and

(c) Greater chance of deterioration and depreciation of inventory, and older stock.

Thus, the optimum ordering quantity must be found out by an appropriate optimization procedure. The discount in price is generally offered in discrete steps, and, hence, the usual calculus method is not applicable. Thus, it will require the help of some special technique to solve the problem. The problem can be considered in the following two forms :

(a) Total or Quantity Discount, and

(b) Incremental Discount.

(a) Total or Quantity Discount

This situation refers to the case, where the supplier gives discount at different slabs but the discount price is applicable to all the units purchased. If C indicates the cost per unit and the amount of purchase is Q, then

$$
\begin{aligned}
C &= c_1 \text{ if } a_1 \leq Q_1 < a_2 \\
&= c_2 \text{ if } a_2 \leq Q_2 < a_3 \\
&\;\;\vdots \\
&= c_n \text{ if } a_n < Q
\end{aligned}
$$

Where $c_1, c_2 \ldots\ldots c_n$ are prices of item as percentage change of C for the corresponding quantities $a_1, a_2, \ldots\ldots a_n$.

Where $c_1 > c_2 > \ldots\ldots > c_n$ . . . (9)

The expression of total cost per unit time can be written as

$$TC(Q_j) = \frac{C_0 D}{Q_j} + \frac{1}{2} C_h Q_j c_j + Dc_j \qquad \ldots (10)$$

$j = 1, 2, \ldots\ldots, n.$

The following procedure can be used to find out quantity ordered $Q = Q^*$ which will minimize TC(Q) as given in (10).

1. Compute the EOQ based on the last value of P(Q), namely c_n, which is the lowest price.

$$Q_n^* = \left(\frac{2C_0D}{c_nC_h}\right)^{1/2} \qquad \ldots (11)$$

If $Q_n^* \geq a_n$, obviously the optimum order quantity is Q_n^*.

2. If $Q_n^* < a_n$, compute the following.

$$Q_{n-1}^* = \left(\frac{2C_0D}{c_{n-1}C_h}\right)^{1/2} \text{ for next lower price} \qquad \ldots (12)$$

Since $c_n < c_{n-1}$

$$Q_{n-1}^* < Q_n^* < a_n \qquad \ldots (13)$$

3. If $a_{n-1} \leq Q_{n-1}^* < a_n$, then TC(Q_{n-1}^*) is compared with TC(a_n).

TC(Q_{n-1}^*) is the minimum value of TC(Q) when C = c_{n-1}, *i.e.*,

$a_{n-1} \leq Q_{n-1} < a_n$. Hence, TC (Q) increases on the right of Q = Q_{n-1}^* when C = c_{n-1}. Therefore, for C = c_{n-1}.

TC(Q) increases of the right of Q = Q_{n-1}^* (see Fig. 1.5).

At Q = a_n, P(Q) = c_n.

The minimum value of TC(Q) when C = c_n occures when Q = Q_n^*. But by equation (13), $Q_n^* < a_n$. Hence for C = c_n, TC(Q) increases on the right of Q_n^* (see Fig. 4). Thus, the optimum order quantity Q is given by

$$Q = \begin{cases} Q_{n-1}^* \text{ if TC } (Q_{n-1}^*) < \text{TC}(a_n) \\ a_n \text{ if TC } (a_n) < \text{TC } (Q_{n-1}^*) \end{cases} \qquad \ldots (14)$$

4. If $Q_{n-1}^* < a_{n-1}$, consider Q_{n-2}^* and follow an approach similar to step-3.

This will be continued until Q_s^* is reached, such that $a_s \leq Q_s^* < a_{s+1}$. The optimum order quantity Q is given by

$$Q = \begin{cases} Q_s^* \text{ if TC } (Q_s^*) < \text{TC}(a_{s+1}) \\ a_{s+1} \text{ if TC } (Q_s^*) > \text{TC } (a_{s+1}) \end{cases} \qquad \ldots (15)$$

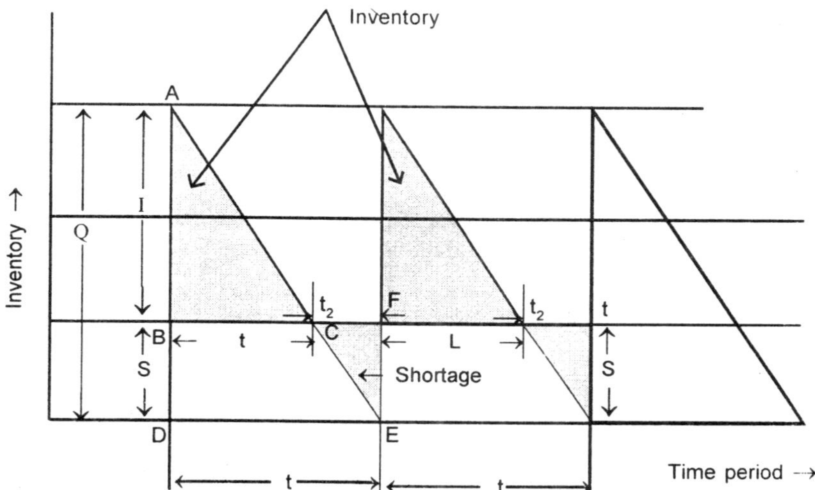

Fig. 4 : EOQ Model for Quantity Discount Quantity.

(b) Incremental Discount

Sometimes discounts are offered only for items which are in excess of the specified amount. These discounts may be offered in different slabs. if Q is the ordering quantity then the cost of each unit of the item is c_1 upto a quantity a_1, the cost of every additional unit beyond a_1 and upto a_2 is c_2. the cost of every additional unit beyond a_2 and upto a_3 is c_3 and so on. It may be assumed that

$$c_1 > c_2 > c_3 \ldots\ldots > c_n$$

The total cost TC(Q) when quantity Q is purchased can be expressed as

$$TC(Q) = c_1 a_1 + c_2(a_2 - a_1) + c_3(a_3 - a_2) + \ldots\ldots + c_i(Q - a_{i-1});\ a_{i-1} < Q \leq a_i$$

$$i = 1, 2, 3, \ldots\ldots n,\quad a_0 = 0.$$

Average cost per item is

$$c(Q) = \frac{TC(Q)}{Q}$$

The expression TC(Q) can be written as

$$TC(Q) = t_i + c_i(Q - a_{i-1});\ a_{i-1} < Q \leq a_i \qquad \ldots (15a)$$

$$= c_1 a_1 + c_2(a_2 - a_1) + \ldots + c_{i-1}(a_{i-1} - a_{i-2}) \qquad \ldots (16)$$

$$t_i = t_{i-1} + c_{i-1}(a_{i-1} - a_{i-2}) \qquad \ldots (17)$$

$i = 2, \ldots n, \ t_1 = 0.$

The cost per unit time F(Q) on lines similar to equation (10) can be obtained as

$$F(Q) = \frac{C_0 D}{Q} + \frac{1}{2} C_h T(Q) + \frac{C_0 TC(Q)}{Q}$$

$$= \frac{D}{Q} [C_0 + TC(Q)] + \frac{1}{2} C_h TC(Q)$$

$$= \frac{D}{Q}[C_0 + t_i + c_i (Q - a_{i-1})] + \frac{1}{2} C_h [t_i + c_i (Q - a_{i-1})]$$

$$= \frac{D}{Q}[C_0 + t_i - c_i a_{i-1}] + \frac{1}{2} C_h C_i Q + Dc_i + \frac{1}{2} C_h t_i - \frac{1}{2} C_h c_i a_{i-1} \quad \ldots(18)$$

Differentiating equation (18) with respect to Q

$$F'(Q) = \frac{D}{Q^2}[C_0 + t_i - c_i a_{i-1}] + \frac{1}{2} C_h c_i \quad \ldots(19)$$

Equating F'(Q) = 0, the solution for Q* is obtained as

$$Q^* = \left[\frac{2D(C_0 + t_i - c_i a_{i-1}}{C_h c_i} \right]^{1/2} \quad \ldots(20)$$

$a_{i-1} < Q^* \leq a_i, \ i = 1, 2, 3, \ldots, n.$

In order to find the value of Q* which will lead to overall minimum optimum Q* for all the intervals $[(a_{i-1} a_i), i = 1, 2, \ldots n]$ are computed. If some of the Q* values computed from equation (1.20) do not belong to the relevant intervals, they are ignored. The values of F(Q*) for all the Q* values which belong to the specified intervals, are then computed. The optimum order quantity is the value of Q* corresponding to which F(Q*) is minimum.

Economic Production Quantity Model (Economic Lot Size Model) (Gradual Supply Case Shortage Not Allowed)

This model is an application of the EOQ concept to the manufacturing environment. EPQ represents the gradual build up of inventory over a period of time as the productdion and consumption go side by side where production rate is higher than the consumption rate. All the assumptions of EOQ model also hold good for this model.

Order size is considered as production size, the annual production rate is taken as P such that P > D, otherwise, if P < D, the item will be used as fast as it is produced (as shown in Fig. 5).

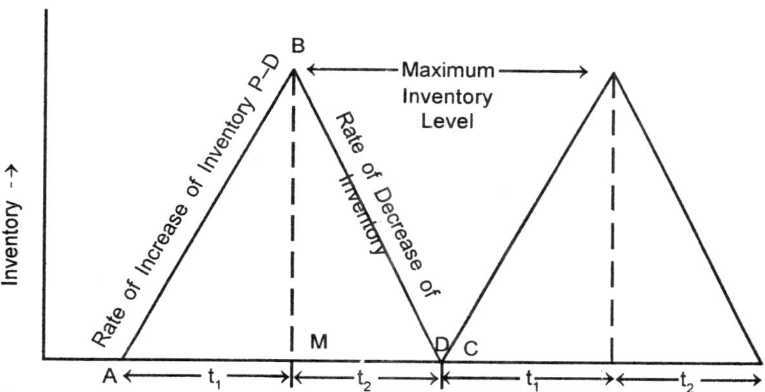

Fig. 5 : Gradual Replacement Inventory Situation (Shortage not allowed).

Cycle time T is the sum of the production time t_1 plus the depletion time t_2 of maximum inventory level BM. Production starts at point A, and stops at point B as soon as the level of inventory becomes BM.

Production time, $t_1 = \dfrac{Q}{P}$

and cycle time, $T = \dfrac{Q}{D}$

Maximum Inventory level BM

$$= (P - D) \times t_1 = (P - D)\frac{Q}{P},$$

Minimum Inventory level = 0

Therefore, average inventory carried

$$= \frac{(P-D)\frac{Q}{P} + 0}{2} = \frac{(P-D)Q}{2P}$$

Total variable cost per year = TC = Annual set up cost + Annual carrying cost

$$= \frac{C_0 D}{Q} + C_h \frac{(P-D)}{P} \cdot \frac{Q}{2}$$

In order to have the optimum value of TC, the dTC/dQ = 0

Thus, $$\frac{dTC}{dQ} = 0 = -\frac{C_0 D}{Q^2} + C_h \frac{(P-D)}{P} \cdot \frac{1}{2} = 0$$

or $$\frac{C_0 D}{Q^2} = \frac{C_h (P-D)}{P} \cdot \frac{1}{2}$$

or $$Q^* = \left[\frac{2C_0DP}{C_h(P-D)}\right]^{1/2}$$

• and Total Variable Cost $$TC^* = \left[\frac{2C_0C_h(P-D)D}{P}\right]^{1/2}.$$

DYNAMIC DEMAND MODELS

In the earlier models, it is assumed that demand is uniform. But in real life situations, the demand may be non-uniform. The demand may be known with certainty, but may vary from one period to the next. Various dynamic models are discussed as follows.

Dynamic Inventory Control (Prescribed Rule Method)

In this method, some rule may be prescribed for procurement of items, *e.g.*, procuring every week, every month or each quarterly.

Dynamic Inventory Model (Fixed EOQ Model)

The average monthly demand = 1500/12 = 125 units per month.

$$\text{Therefore, EOQ} = \left[\frac{2 \times 60 \times 125}{(0.02)(25)}\right]^{1/2} = 173 \text{ approx.}$$

Since full month's requirement is to be ordered, 173 lies between 170 and 280 units. Since 173 is closer to 170 than 280, the order is made for one month requirement at the beginning of January.

Identically, the order is placed for two months' requirement, at the beginning of February.

The details of results are presented in Table 3.

Table 3 : Total Cost Estimation for EOQ Method

Month:	J	F	M	A	M	J	J	A	S	O	N	D	Total
Demand	170	110	80	140	50	80	40	170	140	200	260	60	1500
Replenishment	170	190	–	270	–	–	210	—	140	200	260	60	1500
Starting Inventory	0	0	80	0	130	80	0	170	0	0	0	0	–
Ending Inventory	0	80	0	130	80	0	170	0	0	0	0	0	460

Total ordering cost = 8 × 60 = 480

Total carrying cost = 460 × 0.02 × 25 = 230.

Therefore, the total cost is reduced (from Rs. 895 to Rs. 710) when EOQ policy is adopted instead of three months' rule.

DYNAMIC INVENTORY MODEL(THE SILVER-MEAL HEURISTIC)

In this method, the costs per unit time for the duration of the replenishment quantity are minimized. If the replenishment is done at the beginning of the first period convering requirements upto the end of *T*th period, the average cost per period, A(T) can be computed as following:

$$A(T) = \frac{\text{Ordering cost + Total holding cost to the end of period T}}{T}$$

$$= \frac{C_0 + \text{Holding cost upto T}}{T}$$

If T = 1, order is made for one period and there is no holding cost.

or $$A(1) = \frac{C_0 + 0}{1} = C_0$$

If T = 2, the holding cost will be (period 2 demand carrying for one month x holding cost/unit).

Then $$A(2) = \frac{C_0 + D(2)C_h}{2}$$

Similarly $$A(3) = \frac{C_0 + D(2)C_h + 2D(3)C_h}{3}$$

or $$A(T) = \frac{\left[C_0 + C_h \sum_{j=1}^{j=T} D(j-1)\right]}{T}$$, D(j) is the demand for period j.

PROBABILITY INVENTORY MODEL

This model is the same as discussed with the difference that the stock levels are continuous rather than discrete quantities. Thus, in place of probability $p(D)$, $f(D)dD$ is used, where $f(D)$ is the probability density function of demand. The probability distribution described can be written as,

$$\int_{D=0}^{\infty} f(D)dD = 1 \qquad \ldots (1)$$

Where,

$f(D)$ = continuous probability density function of demand *D*

dD = differential of the demand D

The cumulative probability function is obtained as follows:

$$F(Q) = \int_{D=0}^{\infty} f\,(D)dD \qquad \ldots (2)$$

Showing that F(Q) is the probability of demand lying between *0* and *Q*. For optimal results, the incremental analysis can be used in the same way as has been used in equation (1). The optimal quantity to stock Q^* is the point where

$$F(Q^*) = \frac{C_2}{C_1 + C_2}. \qquad \ldots (3)$$

Multi-Period Probabilistic Models

In single period models, demand is considered to be a major variable factor and the lead time is not supposed to play any role in the decision making. Whereas, in multi-period models, demand as well as lead time are considered in the decision making process. The demand and lead time may be changing according to certain laws of probability, and, thus, there would be risks due to variation in demand and/or in lead time. These risks are reduced by carrying larger inventories, called buffer stocks or safety stocks. However, larger safety stocks lead to more risk, in terms of the funds blocked in inventories, the possibility of obsolescence, and other reasons. Therefore, the objective is to find a rational decision model for balancing the risks of out-of-stock and the risk of inventories, *i.e.,* balancing the carrying cost of the safety stock and the shortage (stock-out) cost.

When there is random variation in demand or lead time, or both, the operating principles are based on inventory control system which calls for establishing the various levels of system's operation, *i.e.,* stock levels. Following kinds of stock levels are fundamentals of the control systems and are fixed for implementing various levels of inventory control system.

1. *Minimum Stock Level (Safety or Buffer Stocks):* This is the minimum stock level allowed to reach, and, thus, as soon as this level is reached, urgent action is taken to bring forward the delivery of the next order. Sometimes, it is known as the danger level. While fixing this level, the main emphasis is laid on the effect of stock-out on the flow of work or operations.

2. *Ordering Stock Level (Reordering Level):* This is the minimum stock level at which ordering action is taken. The major factors to be considered in deciding the re-order level are

   (i) the expected rate of stock consumption, and

(ii) the estimated lead time.

3. *The Hastening Stock Level:* This is the stock level at which hastening action becomes necessary, *i.e.*, requesting suppliers for early delivery. It is subjectively fixéd based on experience between the minimum level and the re-order stock level.
4. *The Maximum Stock Level:* This is the stock level, above which the stock is not allowed to rise and purpose is to curb excess investment. In fixing the maximum stock level, the consideration is mainly financial. The level is so fixed that the value of stock will not become excessive at any time. Other considerations are the possibility of items becoming obsolete, and the danger of deterioration in perishable commodities.

MISCELLANEOUS EXAMPLES

Example 1:

A manufacturer requires an input item at the rate of 2000 pieces per year. The cost of the item is Rs. 50 per piece. The ordering costs have been estimated to be Rs. 25 per order and carrying costs as 20% of average inventory value. Find the following:

(a) Optimal order size,

(b) Number of orders per year,

(c) Time period per order, and

(d) Total cost.

Solution:

Here D = 100,000, $C_0 = 25$, C_h = 20% of average inventory value/unit

$$TC = \frac{250}{Q} + 0.2\,\frac{Q}{2} \quad \text{where Q is order size in rupees}$$

Applying equations (1.2) to (1.5)

$$Q^* = \left(\frac{2 \times 25 \times 100{,}000}{0.2}\right)^{1/2} = 5000$$

$$t^* = \left(\frac{2 \times 25}{0.2 \times 100{,}000}\right)^{1/2} = \frac{1}{20} \text{ year} = 18.25 \text{ days}$$

$$N = \frac{1}{t^*} = 20$$

$$TC^* \text{ (Variable only) } (2 \cdot 25 \times 0.2\ 100{,}00)^{1/2} = 1000.$$

Example 2(a):

If the damand for freeze by a distributor is 6250 units every year, the distributor has no storage space, and, modent thus, the manufacturer has to supply a day's requirement each day. If the manufacturer fails to supply the required numbers of freeze, the shortage cost is Rs. 200 per unit per month. The inventory carrying cost is Rs. 100 per unit per month, and the set up cost per run is Rs. 4000. Compute the optimum run size (Q), the optimum inventory level (I) at the beginning of any period, the optimum scheduling period (t), and the minimum total expected relevant yearly cost (TC).

Solution:

In the given example.

T = 12 months, D = 6250 units.

C_h = Rs. 100 per month,

C_s = Rs. 200 per month,

C_0 = Rs. 4000 per run.

$$Q = \left[\frac{2DC_0(C_h + C_s)}{TC_hC_s}\right]^{1/2}$$

$$= \left[\frac{2 \times 6250 \times 4000\ (100+ 200)}{12 \times 100 \times 200}\right]^{1/2}$$

= 250 units per run.

$$I = \left[\frac{2DC_0C_s}{TC_h(C_h + C_s)}\right]^{1/2}$$

$$= \left[\frac{2 \times 6250 \times 4000 \times 200)}{12 \times 100 \times (100 + 200)}\right]^{1/2}$$

= 167 units.

$$t = \left[\frac{2TC_0(C_h + C_s)}{DC_hC_s}\right]^{1/2}$$

$$= \left[\frac{2 \times 12 \times 4000 \times (100 + 200)}{6250 \times 100 \times 200)}\right]^{1/2}$$

= 2/25 month = 14.4 days.

$$TC^* = \left[\frac{2DTC_oC_s C_h}{(C_h + C_s)}\right]^{1/2}$$

$$= \left[\frac{2 \times 6250 \times 12 \times 100 \times 4000 \times 200)}{100 + 200}\right]^{1/2}$$

$$= \text{Rs. } 200{,}000.$$

Example 2(b):

An item with an annual holding cost of Rs. 5 per unit is replenished about 6 times in a year. The stock-out costs are estimated to be Rs. 3 per week. Consumption rate of the item during lead time of one week per period is given as following:

Consumption rate per week	*200*	*300*	*400*	*500*	*600*
Relative frequency	*0.10*	*0.20*	*0.40*	*0.10*	*0.20*

Record of stock-outs has indicated that the policy of maintaining 400 units as re-order level be reviewed. A decision was taken to retain the same lot size but a safety stock could be carried for the item. Compute the optimal policy.

Solution:

This problem can be solved by following the same general principle as in above case. A zero safety stock is to be associated with a re-order point of 400 units. From the demand pattern, it can be seen that for 400 units of re-order level, stock-outs may be upto 30% of the time. Holding an extra 100 units would limit the stock-out to 20% of the time and 200 units would completely eliminate the chances of stock-out for the same demand pattern. The stock-out costs for the situation requiring 100 units in safety stock, but actually not having a safety stock, would be equal to 100 × 3 × 6 = 1800. These are illustrated in Table (a).

Table (a) : Expected Value of Stock-out Costs

	Safety stock required (in units)			
		Probability		
Safety stock Provided Above 400 Units (in units)	**0 (0.04)**	**100 (0.10)**	**200 (0.20)**	**Expected Value of Stock-out Cost (ESC)**
0	0	100×3×6=1800	200×3×6=3600	Rs. 900
100	0	0	100×3×6=1800	Rs. 360
200	0	0	0	0

The total cost of each of the alternatives is the sum of ESC and the holding costs. For each safety stock increment of 100 units, the annual holding cost is Rs. 100 × 5 = Rs. 500. The total costs are calculated as given in the Table (b).

Table : (b) Total Cost Table

Safety Stock Provided	Expected Stock-out Costs	Expected Holding Costs	Total Costs
0	Rs. 900	0	Rs. 900
100	360	500	860
200	0	1000	1000

From the Table 1.9 it is evident that the original ordering policy should be changed. A safety stock of 100 units be provided and the re-order point should be changed at 500 units.

Example 3:

(a) At present a company is purchasing an item X from outside suppliers The consumption is 10,000 units/year. The cost of the item is Rs. 5 per unit and the ordering cost is estimated to be Rs. 100 per order. The cost of carrying inventory is 25 per cent. If the consumption rate is uniform, determine the economic purchasing quantity.

(b) In the above problem assume that company is going to manufacture the item with the equipment that is estimated to produce 100 units per day. The cost of the unit thus produced is Rs. 3.50 per unit The set-up cost is Rs. 150 per set-up and the inventory carrying charge is 25 per cent. How has your answer changed?

Solution:

(a) From the data of the problem, in using notations, we have

D = 10,000 units per year;

C = Rs. 5 per unit

C_0 = Rs. 100 per order,

C_h = Rs. 25% of Rs. 5 = Rs. 1.25 per year

Economic order quantity is given by

$$= \sqrt{\frac{2 \times 10{,}000 \times 100}{1.25}} = 1{,}265 \text{ units,}$$

(b) Given that

$p =$ 100 units per day;

$C_0 =$ Rs. 150 per set up

$C =$ 3.50 per unit,

$C_h =$ Rs. 25% of Rs. 3.50 = Rs. 0.875 per year.

$d =$ 10,000/250 = 40 units per day (assuming 250 working days in the year)

Economic batch quantity (size) for each production cycle,

$$= \sqrt{\frac{2 \times 10{,}000 \times 150}{0.875}\left(\frac{100}{100-40}\right)}$$

= 2,391 units.

The increase in EBQ in case of (b) may be due to following reasons:

(i) Increased procurement cost (*i.e.,* set-up cost C_0).

(ii) Consumption of inventory simultaneous with production reduces average inventory from Q/2 to {l-(d/p)}.Q/2.

Example 4:

The pie-charts in Fig. represent the relative frequency of lead times and usage rates for an item. The possible lead times are 7, 8, 9, and 10 days with relative frequencies of 0.20, 0.20, 0.30 and 0.30 respectively. The possible demand rates are 15, 30, 40, 50 and 60 units with relative frequency of 0.1, 0.1, 0.2, 0.2, and 0.4, respectively. Compute the optimal policy.

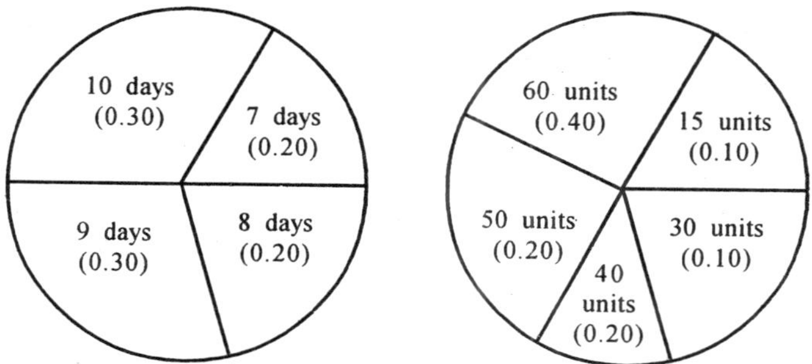

Fig 1 : Relative Frequencies of Lead Times and Demand Rate.

Solution:

Following random numbers are used for the solution:

03991, 38555, 17546, 32643, 69572, 24122, 61196, 30532, 03788, 48228

The simulation of stock movement can be conducted by selecting randomly both a lead time and a demand rate. This can be carried out by mounting spinners on the two pies. The lead time pie will show days, whereas, demand rate pie would point to number of units. The two spins are combined to represent one level of total demand occurring after an order has been placed. A random number table could commonly be used in place of spinners. In a group of 10 numbers, any digit from 0 to 9 is equally likely to occur and, thus, by letting each digit equal to 10%, certain digits can be assigned to each increment of lead time and demand according to its likelihood of occurrence. An arbitrary assignment might be as following:

Days	Lead time Digits	Units Usage Rate per day	Digits
7	0, 1	15	0
8	2, 3	30	1
9	4, 5, 6	40	2, 3
10	7, 8, 9	50	4, 5
		60	6, 7, 8, 9

Then each pair of numbers in the random number table could represent a lead time duration and the number of units demanded per day during that duration. Using the first two numbers of the random numbers, it can be seen that 0 is the first number of 03 and 3 is the second, then the digit 0 corresponds to a lead time of 7 days and demand rate of 40 units per dav. The inventory pattern is, thus, represented in Table.

Table : Simulation of 10 Re-ordering Periods

Random Number	Lead Time	Usage Rate	Demand During Lead Time	Stock on Hand when Order Arrives
0	1	2	3	4
03	7	40	280	+77
38	8	60	480	–123
17	7	60	420	–63
32	8	30	240	+117
69	9	60	540	–183
24	8	50	400	–43
61	9	30	270	+87
30	8	15	120	+237
03	7	40	280	+77
48	9	60	540	–183

From the Table 1.10, the average lead time = 8 days and the average re-order level = 357 units.

Thus, the re-order level is based on an average lead time of 8 days and an average demand rate of 45 units per day. Thus, a typical order is placed when the stock in hand falls to 8 × 45 = 360 units. The right hand column of above Table 1.10 shows that safety stock or buffer stock is not required 50% of the time, while it is required for 50% of the time.

If the simulation procedure is continued, a distribution can be determined for expected stock levels at the time of delivery. From this distribution, an inventory policy can be delivered which establishes a minimum cost balance between holding and opportunity costs. A more direct approach would be to set a tolerable limit for stock-outs per year and hold a safety stock, which confirms to this limit. Most of the time, the safety stock is equal to *K* times, the standard deviation of the demand distribution where K = 1, 2, or 3. Further, it is quite logical that larger number of trials will give simulation closer to actual inventory pattern. However, large number of repeated calculations will require the help of computers for reliable estimate of probable stock-outs.

Example 5(a):

A trader stocks woollen sweaters at the beginning of winter and cannot reorder. The item costs him Rs. 50 each and he sells at Rs. 100. For each sweater that cannot be met on demand, the trader loses a goodwill of Rs. 30. Each unsold piece will have a salvage value of Rs. 20. Holding cost during the period is 10 percent of the item cost. The probability distribution of demand is as follows:

Unit stocked (in hundreds)	*4*	*6*	*8*	*10*	*12*
Demand probability $p(D \leq Q)$	*0.30*	*0.20*	*0.30*	*0.15*	*0.05*

Compute the optimal number of items to be stocked. If a stock level is to be maintained 10 units, find the values of understocking cost (C_2).

Solution:

The data regarding demand distribution given in problem is tabulated in Table.

Table : Probability Distribution of Demand of Woollen Sweaters

Unit stocked (in hundreds)	4	6	8	10	12
Demand probability $p(D = Q)$	0.30	0.20	0.30	0.15	0.05
Cumulative probability $p(D \leq Q)$	0.30	0.50	0.80	0.95	1.00

In the problem, $P = 100, C = 50,$

$C_h = 0.10 \times 50 = 5.0,$

$C_s = 30, S = 20.$

Therefore, $C_1 = C + C_h - S = 50 + 5.0 - 20 = 35$

$$C_2 = P - C - \frac{C_h}{2} + C_s = 100 - 50 - \frac{5.0}{2} + 30 = 77.5$$

$$\text{Thus, } \frac{C_2}{C_1 + C_2} = \frac{77.5}{35 + 77.5} \; 0.69 .$$

Observing the table, this ratio lies between cumulative probabilities of 0.50 and 0.80 which, in turn, reflect the values of Q as 6 and 8. That is,

$$P\,(D \le 6) = 0.50 < 0.69 < 0.80 = P(D \le 8).$$

Therefore, the optimum number of units to stock is 4 units.

Cost of under-stocking can be calculated as follows:

From the problem, following inequality can be found:

$$P(D \le 8) \le \frac{C_2}{35 + C_2} \le P(D \le 10)$$

or
$$0.80 \le \frac{C_2}{35 + C_2} \le 0.95$$

The minimum value of C_2 is determined by letting

$$\frac{C_2}{35 + C_2} = 0.80$$

or
$$C_2 = \frac{(0.80)\,(35)}{(1 - 0.80)} = \text{Rs. } 140$$

The maximum value of C_2 is determined by letting

$$\frac{C_2}{35 + C_2} = 0.95$$

or
$$C_2 = \frac{(0.95)\,(35)}{(1 - 0.95)} = \text{Rs. } 665$$

Therefore, $140 \le C_2 \le 665$.

Inventory for Perishable Products

There are many products, if not sold immediately, have negligible utility, *e.g.*, newspaper, printed programmes for special events, fresh products, etc. Such items are usually associated with large differences between the wholesale cost and the retail price due to high risk involved in stocking the item. He faces obsolescence costs on the one hand and opportunity cost on the other.

Example 5(b):

The following changes in demand.

The average daily demand of the item is 100 units per day and variance of daily demand is (80 units/day)^2.

Solution:

Average lead time consumption,

$U^* = d^* t^*_L = 100 \times 15 = 1500$ units

$\text{Var}(U) = (\text{Var } d)t^*_L + (\text{Var } t_L)(d^*)^2$

$= 80 \times 15 + 46 \times (100)^2 = 461{,}200$

Therefore, $\sigma_U = [\text{Var}(U)]^{1/2}$

$= 679$ units per lead time

Desired service level is 94.5%, therefore

Buffer stock $= Z\sigma_{U*} = (1.60)(679) = 1086$

and Reorder level $= R^* = 1500 + 1086$

$= 2586$ units.

Example 5(c):

A newspaper man buys papers for Rs. 0.90 each and sells them for Rs. 1.50 each. He cannot return unsold newspapers. Daily demand has the following distribution:

Number of Customers	*250*	*260*	*270*	*280*	*290*	*300*	*310*	*320*	*330*	*340*
Probability	*0.01*	*0.03*	*0.06*	*0.10*	*0.20*	*0.25*	*0.15*	*0.10*	*0.05*	*0.05*

If each day's demand is independent of the previous day's demand, how many papers should he order each day?

Solution:

The given data can be transferred in the Table.

Table : Probability Distribution of Demand

Number of Customers	250	260	270	280	290	300	310	320	330	340
Probability	0.01	0.03	0.06	0.10	0.20	0.25	0.15	0.10	0.05	0.05
Cumulative Probability	0.01	0.04	0.10	0.20	0.40	0.65	0.80	0.90	0.95	1.00

$$\text{The ratio, } \frac{C_2}{C_1+C_2} = \frac{0.60}{0.90+0.60} = \frac{0.60}{1.50} = 0.40$$

Thus the value of Q such that

$$P(D \le Q^* - 1) \le 0.40 \le P(D \le Q^*),$$

gives $Q^* = 300$. Thus, newspaper man should buy 300 papers each day.

Example 6:

A firm orders an item 6 times a year with a specified service level of one stock out per 3 years. Daily demand of the item is 100 units. The history of re-order lead times is given below. Compute the re-order level.

Past Record of Lead Times (1998) (Date/Month)

Order placed	*17/1*	*13/2*	*26/3*	*6/4*	*12/5*	*12/6*
Order received	*28/1*	*28/2*	*30/4*	*28/4*	*30/5*	*28/6*
Lead time (Days)	*11*	*15*	*35*	*22*	*18*	*16*
Working days	*7*	*10*	*28*	*18*	*15*	*12*

Solution:

$$\text{Average Lead Time} = t^*_L = \frac{7+10+28+18+15+12}{6} = 15 \text{ days}.$$

Variance of lead time

$$= \frac{(7-15)^2 + (10-15)^2 + \ldots\ldots + (12-15)^2}{6} = 46 \text{ days}^2$$

$$U^* = dt_L = 100(15) = 1500 \text{ units demanded per lead time}$$

Variance $U = 46\,(100)^2$

or $\sigma_u = 100(46)^{1/2} = 680$

Desired SOR per year

$$= \frac{\text{Average number of Stock} - \text{out per year}}{\text{Number of cycles per year}} = \frac{1/3}{6}$$

$$= 0.055 = 5.5\%$$

$= $ Probability of stock-out per order cycle.

Thus, from normal table, for service level of 100% – 5.5% = 94.5%, the value of *Z* from the normal distribution table is 1.60

Therefore, safety stock (Buffer stock) = 1.60×680 = 1088 units, and

Reorder level $= R^* = U^* + Z\sigma_U$

= 1500 + 680 = 2180 units.

Example 7:

A company has to purchase four items A, B, C and D for the next year. The projected demand and unit price (in Rs.) are as follows:

Item	*Demand (units)*	*Unit Price (Rs.)*
A	60,000	3
B	40,000	2
C	1,200	24
D	5,000	4

If the company wants to restrict the total number of orders to 40 for all the four items, how many orders should be placed for each item?

Solution:

The computation of the allocation of 40 total orders per year to four items is shown below:

Items	**Demand (units)**	**Unit Price (Rs.)**	**√DC**	**√DS/Σ.√DC**	**Number of Orders per Year**
A	60,000	3	424.26	0.416	0.416 (40) = 16.64 = 17
B	40,000	2	282.84	0.277	0.227 (40) = 11.08 = 11
C	1,200	24	169.70	0.166	0.166 (40) = 6.64 = 7
D	5,000	4	141.42	0.138	0.138 (40) = 5.52 = 5
			1,018.22		40

Example 8(a):

A manufacturing firm has a fixed weekly cyclic demand as follows:

	Mon	*Tue*	*Wed*	*Thur*	*Fri*	*Sat*	*Sun*
Demand in units	*10*	*18*	*3*	*1*	*20*	*10*	*15*

Daily production is to be maintained all the seven days of a week. Shortage cost is Rs. 5 per unit per day and carrying cost depends upon the size of the quantity carried (Q), as follows:

Cost per unit for one day (Rs.) =	*2*	*5*	*8*
Q	$1 \le Q \le 4$	$5 \le Q \le 15$	$15 \le Q$.

The charges are based on the situation at the end of the day. Compute the optimal starting stock level.

Solution:

Let the production rate be the average of the total sales which is

$$\frac{10+18+3+1+20+10+15}{7} = 11 \text{ units per day}$$

The total weekly cost can be computed by tabulation for various starting stocks. Tables (a) to (b) are for starting stocks of 9, 10, and 11 units respectively.

Table (a) : Cost Analysis with 9 Initial Stock

Days	Onhand Stock Start of Day	Demand	Inventory	Shortage Cost (Rs.)	Carrying Cost (Rs.)
Mon	9	10	–1	5	0
Tue	–1 + 11 = 10	18	–8	40	0
Wed	–8 + 11 = 3	3	0	0	0
Thur	0 + 11 = 11	1	10	0	50
Fri	10 + 11 = 21	20	1	0	2
Sat	1 + 11 = 12	10	2	0	4
Sun	2 + 11 = 13	15	–2	10	0
Mon	–2 + 11 = 9	Total cost		55 +	56 = 111

Table (b) : Cost Analysis with 10 Initial Stock

Days	Onhand Stock Start of Day	Demand	Inventory	Shortage Cost (Rs.)	Carrying Cost (Rs.)
Mon	10	10	0	0	0
Tue	0 + 11 = 11	18	–7	35	0
Wed	–7 + 11 = 4	3	1	0	2
Thur	1 + 11 = 12	1	11	0	55
Fri	11 + 11 = 22	20	2	0	4
Sat	2 + 11 = 13	10	3	0	6
Sun	3 + 11 = 14	15	–1	5	0
Mon	–1 + 11 = 10	Total cost		40 +	67 = 107

Table (c) : Cost Analysis with 11 Initial Stock

Days	Onhand Stock Start of Day	Demand	Inventory	Shortage Cost (Rs.)	Carrying Cost (Rs.)
Mon	11	10	1	0	2
Tue	1 + 11 = 12	18	–6	30	0
Wed	–6 + 11 = 5	3	2	0	4
Thur	2 + 11 = 13	1	12	0	60
Fri	12 + 11 = 23	20	3	0	6
Sat	3 + 11 = 14	10	4	0	8
Sun	4 + 11 = 15	15	0	0	0
Mon	0 + 11 = 11	Total cost		30 +	80 = 110

Thus, the optimal solution is to start with 10 units as starting stock on Monday for which the minimum total cost would be Rs. 107.

Procedure for solving such problem is incremental cost analayis which is self explanatory through this example.

Example 8(b):

An *X-Y* manufacturing Company purchases bush bearings for fans having an annual demand of 25,000 units. The ordering cost is Rs. 500, inventory carrying cost is 20% of the purchase price per year. The purchase prices are:

$c_1 =$ Rs. 12 for $Q_1 < 4000$

$c_2 =$ Rs. 10 for $4000 \leq Q_2 < 5000$

$c_3 =$ Rs. 8 for $5000 \leq Q_3$

Compute the optimum purchase quantity.

Solution:

As per step-1, $$Q_3^* = \left[\frac{2 \times 500 \times 25{,}000}{0.2 \times 8}\right]^{1/2} = 3953$$

Since 3953 < 5000, next computation is performed, *i.e.*,

$$Q_2^* = \left[\frac{2 \times 500 \times 25{,}000}{0.2 \times 10}\right]^{1/2} = 3535$$

Since 3535 < 4000, next computation is performed, *i.e.*,

$$Q_1^* = \left[\frac{2 \times 500 \times 25{,}000}{0.2 \times 12}\right]^{1/2} = 3227$$

$$= 3227 > 0.$$

Now the total costs for purchase

$Q_1^* = 3227$, a_2 4000,

and $a_3 = 5000$ units are compared.

TC_1 (for purchasing 3227 units)

$$= 12 \times 25000 + \frac{500 \times 25{,}000}{3227} + \frac{1}{2}(0.2) \times 12 \times 3227$$

$= 307{,}746$

TC_2 (for purchasing 4000 units)

$$= 10 \times 25000 + \frac{500 \times 25{,}000}{4000} + \frac{1}{2}(0.2) \times 10 \times 4000$$

$= 257{,}125$

TC_3 (for purchasing 5000 units)

$$= 18 \times 25000 + \frac{500 \times 25{,}000}{5000} + \frac{1}{2}(0.2) \times 8 \times 5000$$

$= 206{,}500$*

Therefore, the economic puchase order for the given problem is

$Q_3^* = 5000$ units.

Example 9:

The demand of an item by a firm is estimated as follows:

Month:	*J*	*F*	*M*	*A*	*M*	*J*	*J*	*A*	*S*	*O*	*N*	*D*	*Total*
Demand:	*150*	*90*	*60*	*120*	*20*	*60*	*20*	*150*	*120*	*180*	*240*	*40*	*1250*

Ordering cost is Rs. 60 and the carrying cost per unit per month at the end of each month is 2%, unit Cost is Rs. 20. Supply is instantaneous, and thus, there is no lead time and no stock-outs. Further, requirement ordered is for full month. Compute the total annual cost for three months policy.

Solution:

Total cost for the policy of ordering quarterly can be computed as shown in Table.

Table : Total Cost Estimation for Three Months Policy

Month:	J	F	M	A	M	J	J	A	S	O	N	D	Total
Demand	170	110	80	140	50	80	40	170	140	200	260	60	1500
Replenishment	360	–	–	270	–	–	350	—	–	520	–	–	1500
Starting Inventory	0	190	80	0	130	80	0	310	140	0	320	60	–
Ending Inventory	190	80	0	130	80	0	310	140	0	320	60	0	1310

Total replenishment cost = 4 × 60 = Rs. 240.

Total carrying cost = 1310 × (0.02) (25) = Rs. 655.

Total annual cost = 240 + 655 = Rs. 895.

Example 10:

A forged automobile part is used at the rate of 200 per day and can be manufactured at a rate of 1200 per day. It costs Rs. 4000 to set up the manufacturing process and Rs. 0.2 per unit per day held in inventory based on the actual inventory time. If the shortage is not allowed, compute the minimum cost and the optimum number of units per manufacturing run.

Solution:

Here D = 200 units, P = 1200 units,

C_0 = Rs. 4000, C_h = Rs. 0.20

$$\text{Therefore, } Q^* = \left[\frac{2 \times 4000 \times 200 \times 1200}{(0.2)\,(1200-200)}\right]^{1/2} = 3098 \text{ units}$$

and

$$TC^* = \left[\frac{2 \times (0.2) \times 4000 \times (1200-200) \times 200}{1200}\right]$$

= Rs. 516.3.

Example 11:

The daily demand of sugar is normally distributed with a mean of 400 units per day and standard deviation of 10 units. Lead time is 5 days. The cost of placing an order is Rs. 100, and annual holding costs are 20% of the unit price of Rs. 10. A 95% level is desired. Back orders are allowed, but there is no stock-out cost. Find the various levels.

Solution:

$$Q^* = \left(\frac{2C_0D}{C_h}\right)^{1/2} = \left(\frac{2\times 400\times 365\times 100}{0.2\times 10}\right)^{1/2} = 3824$$

From the normal distribution, a 95% confidence level gives Z = 1.645

Thus $R^* = U^* + S = d^* t_L + Z\sigma_U = 400(5) + 1.645\sigma_U$

where $\sigma_U = (t_L\sigma^2_d)^{1/2} = (5 \times 10^2)^{1/2} = 22.4$

Therefore safety stock, S = 1.645 (22.4) = 37 units

and $R^* = 400(5) + 1.645(22.4) = 2037$ units.

Operating rule would be to order 3828 units when an order point of 2037 units is reached.

Average inventory level = 3828/2 + 37 = 1951 units

Maximum inventory level = 3828 + 37 = 3865 units.

Example 12:

The following observation for 7 days. All other cost factors are the same.

Actual daily demand	*301-500*	*501-700*	*701-900*	*901-1100*
Number of occurrence(Days)	*3*	*2*	*1*	*1*
Relative frequency of occurrencc in percent	*42.8*	*28.6*	*14.3*	*14.3*

Solution:

$$\text{Expected Demand} = d^* = \frac{400\times 3 + 600\times 2 + 800\times 1 + 1000\times 1}{7}$$

$$= 600 \text{ units per day}$$

$$\sigma_d = \frac{400\times 3 + 600\times 2 + 800\times 1 + 1000\times 1}{7}$$

$$= 600 \text{ units per day}$$

$$= \left[\frac{3(400-600)^2 + 2(600-600)^2 + 1(800-600)^2 + 1(1000-600)^2}{7}\right]^{1/2}$$

$$= 214$$

Therefore $U^* = d^*_L = 600 \times 5 = 3000$ units.

$$\sigma_U = \left[\frac{5(32,00,00}{7}\right]^{1/2} = 486$$

Therefore, $R^* = 600 \times 5 + 1.645(486) = 3800$ units

Safety stock S = 800 units

$$Q^* = \left[\frac{2 \times (600)\,(365)\,(100)}{(0.2)(10)}\right]^{1/2} = 4770 \text{ units}.$$

Operating rule would be to order 4770 units when inventory level reaches to the reorder point of 3800 units.

Maximum level = 4770 + 800 = 5570 units.

Example 13:

A trader sells an item for Rs. 50 per unit. The item costs Rs. 30. Unsold items can be sold for Rs. 10 each. There is no shortage penalty cost besides the lost revenue. The demand is known to be any value between 500 and 800 items. Find the optimal number of units of the item to be stocked.

Solution:

In the present problem, it is assumed that the demand distribution is a uniform distribution between 500 and 800 units represented by the relation

$$f(D) = \frac{1}{a - b}$$

Where b is the upper limit of the demand and a is the lower limit. Thus, for the present problem,

$$f(D) = \frac{1}{800 - 500} = \frac{1}{300}$$

for $500 \le D \le 800$.

Other factors are as follows:

Selling price = P = Rs. 50,

Purchase price = C = Rs. 30,

Salvage value = S = Rs. 10,

Carrying cost = 0,

Shortage cost = 0

Therefore, $C_1 = C + C_h - S = 30 + 0 - 10 =$ Rs. 20

$$C_2 = P - C - C_h/2 + C_s = 50 - 30 - 0 - 0 = \text{Rs. } 20.$$

Then $$\frac{C_2}{C_1 + C_2} = \frac{20}{20+20} = \frac{20}{40} = 0.5$$

Now using equations (1.29) and (1.30).

$$\int_{D=500}^{Q} f(D) = \int \frac{dD}{300} = \frac{Q}{300} - \frac{500}{300}$$

or $$\frac{Q}{300} - \frac{5}{3} = 0.5$$

or $$Q = 650 \text{ units.}$$

Hence, the optimal number of units to be stocked is $Q^* = 650$.

Example 14:

A product is sold at the rate of 50 pieces per day and is manufactured at a rate of 250 pieces per day. The set-up cost of the machines is Rs. 1,000 and the storage cost is found to be Rs. 0.0015 per piece per day.

With labour charges of Rs 3.20 per piece, material cost at Rs. 2.10 per piece and overhead cost of Rs. 4.10 per piece, find the minimum cost batch size if the interest charges are 8 per cent (assume 300 working days in a year). Compute the optimal number of cycles required in a year for the manufacture of this product.

Solution:

From the data of the problem, in usual notations, we have

$D = 50 \times 300 = 15{,}000$ pieces per year;

C_y = Rs. 100 per production run

$p = 250 \times 300 = 75{,}000$ pieces per year (production rate)

$$C_h = 0.0015 \times 300 + 0.08(3.20 + 2.10 + 4.10)$$

$$= \text{Rs. } 1.202 \text{ per year.}$$

Economic batch size for each production cycle

$$= \sqrt{\frac{2 \times 15{,}000 \times 1{,}000}{1.202}\left(\frac{75{,}000}{75{,}000 - 15{,}000}\right)}$$

$= 5{,}586$ pieces (approx.)

Optimal number of cycles = 3 cycles (approx.).

Example 15:

A forget item is used in an automobile industry at a uniform rate of 1250 units per year. Each item costs Rs. 60. Ordering and Inspection cost is Rs. 75 per order. Holding costs are 20% of the value of the average inventory. The distribution of lead time is as follows:

Lead time days	*7*	*8*	*9*	*10*	*11*	*12*	*13*	*14*	*15*
Relative frequency	*0.00*	*0.04*	*0.08*	*0.38*	*0.24*	*0.12*	*0.09*	*0.03*	*0.02*

The stock-out cost is estimated to be Rs. 50 per unit. The number of working days in a year are 250. Determine the optimum stock policy.

Solution:

In the given problem,

C_0 = Rs. 75, $C_h = 60 \times 0.2$ = Rs. 12,

D = 1250 units

Therefore

$$Q^* = \left[\frac{2 \times 75 \times 1250}{12}\right]^{1/2} = 125 \text{ units}$$

Orders per year = 1250/125 = 10.

Time between orders = 250/10 =25 days.

and a daily usage rate = 1250/250 = 5 units per day

The average lead time is obtained as

$$8 \times 0.04 + 9 \times 0.08 + 10 \times 0.38 + 11 \times 0.24 + 12 \times 0.12 + 32 \times 0.09 + 14 \times 0.03 + 15 \times 0.02 = 10.81 \text{ days.}$$

Standard deviation of lead time

$$\sigma_{t_L} = [(10.81 - 4)^2(0.04) + (10.81 - 9)^2(0.08) + \ldots + (15 - 10.81)^2(0.02)]^{1/2} = 1.38.$$

For 98% confidence level, the lead time

$$= t^*_L + Z\sigma_{t_L} = 10.81 + 2.06\,(1.38) = 13.65 \text{ days.}$$

Now, the calculation is made for the expected costs of lead time alternatives of 12, 13, 14 and 15 days respectively. Since, the opportunity cost is Rs. 50 per unit, *i.e.,* a loss of Rs. 50 × 5 = Rs. 250 per day can be sustained for late deliveries based on the daily usage of 5 units. Thus a table is developed for expected value multiplying the stock deficiencies by Rs. 50 per unit and the excess stock by holding cost of Rs. 12 per unit.

Table : Expected Costs of Lead Time Alternatives

Lead Time Provided	Lead time required in days Figures in brackets are probabilities								Expected Holding cost (EHC)	Expected Stock-out Cost (ESC)
	8 (0.04)	9 (0.08)	10 (0.38)	11 (0.24)	12 (0.12)	13 (0.09)	14 (0.03)	15 (0.02)		
12	240	180	120	60	0	250	500	750	84	52.5
13	300	240	180	120	60	0	250	500	135.6	17.5
14	360	300	240	180	120	60	0	250	192.6	5
15	420	360	300	240	180	120	60	0	251.4	0

Figures to the left of zeros are holding costs and that to the right of zeros are stock-out costs.

The expected holding costs and expected stock-out costs affect the ordering policy in following two ways:

(a) Holding costs are annual expenses, and, thus, reflect the extra inventory or safety stock held in storage.

(b) Stock-out costs occur only when the arrival of an order takes more time than the lead time allowed. This can be taken as ordering expenses or set-up expenses.

These are calculated as follows:

For the lead time alternative of 12 days.

$(EHC)_{12}$ = $240 \times 0.04 + 180 \times 0.08 + 120 \times 0.38 + 60 \times 0.4$ = Rs. 84

= The expected holding cost of safety stock for 12 days lead time.

and

$(ESC)_{12}$ = $250 \times 0.09 + 500 \times 0.03 + 750 \times 0.02$ = Rs. 52.5

= Expected stock-out costs for 12 days' lead time and EOQ for 12 days lead time is given as

$$Q_{12} = \left[\frac{2[C_0 + (ESC)_{12}]D}{C_h}\right]^{1/2} = \left[\frac{2(75) + 52.5(1250)}{12}\right]^{1/2} = 163$$

and total costs for 12 days lead time

$TC_{12} = [75 + 52.5) \times 1250]/163 + (12 \times 163)/2 + 84$

= Rs. 1980.

All these are summarized in Table.

Table : Optimal Results

Lead Time Provided (Days)	Economic Order Quantity	Total Cost
12	163	1980
13	138	1802
14	129	1742
15	125	1751

From Table 1.6, the most economic policy is to allow 14 days lead time with EOQ = 129 units with least total cost of Rs. 1742. An order should be placed whenever the stock level falls to 14 × 5 = 70 units. By following this policy one can expect that run-out of stock would be twice in 100 order periods.

Example 16:

The annual requirements of a company for steel bar is 80 tonnes. The ordering cost per order is Rs. 2000, and the inventory carrying cost is 20% of average inventory per year. The cost per tonne of steel is Rs. 15,000. Find the economic order quantity.

Solution:

Various calculations are shown in Table.

Table : Tabular Approach to Economic Order Quantity

Order per Year	Lot Size	Average Inventory	Carrying cost 20% per Year	Ordering Cost	Total Cost per Year
1	80	40	120,000	2000	122,000
2	40	20	60,000	4000	64,000
4	20	10	30,000	8000	38,000
6	13.33	6.67	20,000	12,000	32,000
8	10*	5	15,000	16,000	31,000*
10	8	4	12,000	20,000	32,000
12	6.67	3.33	10,000	24,000	34,000
16	5	2.5	7,500	32,000	39,500
24	3.33	1.67	5,000	48,000	53,000
32	2.5	1.25	3,750	64,000	67,750

* Optimal situation for EOQ

It can be observed that an order size of 10 tonnes will result in the lowest total cost over the year of 10 alternatives studied. But this may not be the exact solution, for which more studies would be required in between lots of 13.33 and 8 tonnes. In fact, the situation corresponds to the point where ordering cost is equal to carrying cost. This can be found out by further studying the total costs in between 13.33 and 8 tonnes.

EXERCISES

1. An item is produced at the rate of 50 items per day. The demand occurs at the rate of 25 items per day. If the set-up cost is Rs. 100 per set-up and holding cost is Re 0.01 unit of item per day, find the economic lot size for one run, assuming that the shortage are not permitted. Also find the time of cycle and minimum total cost for one run.
2. (a) What are the various assumptions of EOQ formula?

   (b) Identify the two basic decisions addressed by inventory management and discuss why the responses to these differ for continuous and periodic inventory systems.
3. A retail store sells 5,200 units of a product in a year. Each unit costs Rs. 2 to the store. The wholesaler charges Rs. 10. Charges on the working capital are 15 per cent and the insurance charges on inventory amount to 5 per cent per annum. All other expenses are either fixed in nature or do not vary with the 'level of inventory or the quantity ordered. The owner is presently following the policy of ordering 100 units every week. He wishes to evaluate his inventory policy. What recommendation would you make?
4. Explain in detail, what constitute the ordering cost and carrying cost?
5. What are inventory models? Enumerate various types of inventory models and describe them briefly.
6. Discuss the various costs involved in an inventory model.
7. Derive an EOQ formula with different rates of demand in different cycles.
8. Explain clearly the various costs that are involved in inventory analysis with suitable examples. How are they inter-related?
9. What are the categories of costs that are associated within developing a sound inventory model? What are the components of costs under each of them?

10. Amit manufactures 50,000 bottles of tomato ketchup in an year. The factory cost per bottle is Rs. 6, the set-up cost per production run is estimated to be Rs. 90, and the carrying costs on finished goods inventory amount to 20 per cent of the cost per annum. The production rate is 600 bottles per day, and sales amount to 150 bottles per day. What is the optimal production lot size and the number of production runs?

11. Assuming you are reviewing the production lot size decision associated with a production operation where the production rate is 8,000 units a year, annual demand is 2000 units, set-up cost is Rs. 300 per production run and holding cost is Rs. 1.60 per unit per year. The current production run is 500 units every 3 months.

    Would you recommend a change in the production lot size? If so, why? How much could be saved by adopting the new production run lot size?

12. A company uses annually 24,000 units of raw material which costs Rs. 1.25 per unit. Placing each order costs Rs. 22.50 and the carrying cost is 5.4 per cent per year of the average inventory. Find the economic lot size and the total inventory cost (including the cost of material).

13. A factory requires 1,500 units of an item per month, each costing Rs. 27. The cost per order is Rs. 150 and the inventory carrying charges work out to 20 per cent of the average inventory. Find out the'economic order quantity and the number of orders per year. Would you accept a 2 per cent price discount on a minimum supply quantity of 1,200 units? Compare the total costs in both the cases.

14. Annual demand for a particular item of inventory is 10,000 units. Inventory carrying cost per unit per year is 20 per cent and ordering cost is Rs. 40 per order. The price quoted by the supplier is Rs. 4 per unit. However the supplier is willing to give discount of 5 per cent for orders of 1,500 or more. Is it worthwhile to avail of the discount offer?

15. A materials manager has the following data for procuring a particular item. Annual demand = 1,000. Ordering cost = Rs. 800. Inventory carrying cost = 40 per cent. Cost per item = Rs. 60. If the order quantity is more than or equal to 300, a discount of 10 per cent is given. For how much should he place the order in order to minimize the total variable cost?

16. (a) The soft goods department of a large department store sells 500 units per month of a certain large bath towel. The unit cost of

a towel to the store is Rs. 10 and the cost of placing an order has been estimated to be Rs. 50. The store uses an inventory carrying charge of 20 per cent of average inventory valuation per month. Assuming that the demand is deterministic and continuous, and that no stock outs are allowed determine the optimal order quantity. What is the time between placing of orders? The procurement lead time for the towels is one month. What is the reorder point based on the on-hand inventory level?

(b) Consider an item on which incremental quantity discounts are available. The first hundred units cost Rs. 100 each and additional units cost Rs. 95 each. For this item, demand = 500 units per year, inventory carrying charge = 20 per cent of average inventory valuation per annum, and procurement cost = Rs. 50. Determine the EOQ.

17. A manufacturing concern requires 2,000 units of a material per year. The ordering costs are Rs. 10 per order, while carrying costs are Re 0.16 per year per unit of average inventory. The purchase price is Re 1 per unit. Find the economic order quantity, and the total inventory cost. If a discount of 5 per cent is available for orders of 1,000 units, should the manufacturer accept" this offer? Also, if he purchases a single lot of 2,000 units, he has to pay Re 0.93 per unit. What purchase quantity would you recommend?

18. The annual demand for a product is 64,000 units (or 1,230 units per week). The buying cost per order is Rs. 10 and the estimated cost of carrying one unit in stock for a year is 20 per cent. The normal price of the product is Rs. 10 per unit. However, the supplier offers a quantity discount of 2 per cent on an order of at least 1,000 units at a time, and a discount of 5 per cent if the order is for at least 5,000 units. Suggest the most economic purchase quantity per order.

19. (a) Discuss in brief: (i) re-order level, (ii) reserve stock, (iii) lead time.

(b) What is lead time? What activities occur during lead-time? What bearing does this have on reserve stock?

20. Explain the meaning of inventory control. Give three methods of classification techniques and their uses for inventory control.

21. What is economic order quantity (EOQ) concept? Explain the basic ideas involved in this concept.

22. Explain the concept of the Q-system, the P-system and the Two-Bin system for management of inventories by giving appropriate examples.
23. Differentiate between a fixed-order quantity system and a fixed-interval system. Mention the advantages and disadvantages of both.
24. Explain the problem of inventory control with deterministic demand.
25. For a certain item the demand rate is 1,000 units/year. The ordering cost per order is Rs. 10; cost of the item Re 0.50 and the inventory carrying charges are 30 per cent per year. The lead time can be assumed as 2 years. Determine (i) optimal order quantity; (ii) re-order point; and (iii) minimum average cost.
26. A wholesaler supplies 30 stuffed dolls each week day to various shops. Dolls are purchased from the manufacturer into lots of 120 each for Rs. 1,200 per lot. Every order incurs a handling charge ofRs 60 plus to freight charge of Rs. 250 per lot. Multiple and fractional lots also can be ordered, and all orders are filed the next day. The incremental cost is Rs. 0.60 per year to store a doll inventory. The wholesaler finances inventory investments by paying its holding company 2 per cent monthly for borrowed funds. How many dolls should be ordered for at a time in order to minimize the total annual inventory cost? Assume that there are 250 weekdays in a year. How frequently should he order?
27. Derive the rule that gives optimum order quantity for a single period stochastic inventory system for which holding cost and shortage cost are proportional to time and quantity. Assume that the demand is discrete.
28. For the following inventory problem find out (i) how much should be ordered each time? (ii) When should the order be placed? (iii) What should be the inventory level (ideally) immediately before the material ordered is received?

Annual demand	= 7,200 units (360 days)
Cost per unit	= Re 1
Ordering cost/order	= Rs. 12/order
Inventory carrying charge	= 24 per cent
Normal lead time	= 15 days
Safety stock	= 30 days consumption.

29. An automobile company has determined that 16 spare engines will result into a stockout risk of 15 per cent while 20 will reduce the risk

to 15 per cent and 24 to 10 per cent. If the lead time is 3 months and the average usage is 6 engines per month, what should be the re-order level to maintain 85 per cent service level?

30. The average demand for an item is 120 units per year. The lead time is one month and the demand during lead time follows normal distribution with average of 10 units and standard deviation of 2 units. If the item is ordered once in 4 months and the policy of the company is that there should not be more than one stock-out every two years, determine the re-order lead time.

31. A company uses annually 50,000 units of an item, each costing Rs. 1.20. Each order costs Rs. 45 and inventory carrying costs 15 per cent of the annual average inventory value.
    (i) Find EOQ.
    (ii) If the company operates 250 days a year, the procurement time is 10 days and buffer stock is 500 units, find the reorder level, maximum, minimum and average inventory.

32. The probability distribution of demand of an item is as follows:

Monthly demand :	0	1	2	3	4	5
Probability :	0.1	0.2	0.2	0.3	0.1	0.1

The cost of carrying inventory is Re 1 per unit per month. The current policy is to maintain a stock of three items at the beginning of each month. Assuming that this level is the optimum level, then what is the shortage cost of one item for one time unit.

33. A fish stall sells a variety of fish at the rate of Rs. 5 per kg on the day of the catch. If the stall fails to sell the catch on the same day, it pays for storage at the rate of Re 0.30 per kg and the price fetched is Rs. 4.50 per kg on the next day. Past records show that there is an unlimited demand for fish one day old. The problem is to ascertain how much fish should be procured everyday so that the total expected cost is minimum. It has been found from the past records that the daily demand follows an exponential distribution with,

$$f(x) = 0.02\ xe^{-0.02},\quad 0 < x \leq \infty.$$

34. Discuss a probabilistic reorder point lot size inventory model to determine the optimal reorder point for a prescribed lot size. Demand is random and continuous with given probability density function. Shortages are not allowed. Lead time is zero. Usual notations may be used.

35. Consider the single period model with C_h = Re 1; C_s = Rs. 19 and 1 = 1.12 units. The demand density function is given by f(D) = De^{-D}. Find the optimum value of Q*.

36. For a fixed order quantity system, find out (i) economic order quantity, (ii) Optimum buffer stock, (iii) Re-order level for an item with the following data:

Annual consumption	= 10,000 units,
Cost of one unit	= Re 1.00,
Set-up cost	= Rs. 12 per production run,
Carrying cost	= Re 0.24 %.

Past lead times: 15 days, 25 days, 13 days, 14 days, 30 days, 17 days.

37. A company uses annually 24,000 units of a raw material which costs Rs. 1.25 per unit. Placing each order costs Rs. 22.5 and the carrying cost is 5.4 per cent of the average inventory. Find the economic order quantity, and the total inventory cost (including the cost of material). Should the company accept the offer made by the supplier of a discount of 5 per cent on the cost price on a single order of 24,000 units?

    Assume the company works for 300 days in a year. If the procurement time is 12 days and buffer stock is 400 units, find the re-order point, the minimum, maximum and average inventory.

38. (a) Explain ABC analysis. What are its advantages and limitations, if any?

    (b) Describe the norms you would use for controlling inventories classified by ABC analysis.

39. Explain the importance of ABC analysis in the problem of inventory control of an organization using a large number of items.

40. Explain the basis of selective inventory control and state the different selection techniques adopted in inventory control system. Give brief note on each.

41. What is ABC analysis? For what purpose do the inventory managers use ABC analysis? Explain the use of ABC analysis to various functional areas.

42. "Purchase manager should shoulder special responsibility for A-items and the 'A' items should not be handled on any routine procurement policy." Discuss.

43. What are the objectives of inventory control? Describe the method of carrying out ABC analysis and suggest a percentage of items, their consumption value and financial limits prescribed for ABC category in a typical organization.
44. 'ABC analysis is a very useful approach for selective inventory control but has some major limitations.' Do you agree with this statement? Explain how these limitations, if any, can be removed.
45. What are the advantages and disadvantages of increased inventory? Briefly explain the objectives that must be fulfilled by an inventory control system.
46. Why should the order size in case of gradual receipt of goods be larger than the one in case of instantaneous receipt?
47. With the help of a Quantity Cost Curve, explain the significance of economic order quantity. What are the limitations in using the formula for EOQ?
48. It is said that EOQ models however complex are restricted by so many assumptions that they have very limited practical value. Do you agree with this view? Illustrate your answer with examples.
49. Control of inventory (stock) is an important aspect of management. You are required to discuss:
    (a) the purpose of inventory, and
    (b) method to ensure that the optimal investment is made in this asset.
50. Derive the optimal economic lot size formula in the usual notations when the rate of replenishment is finite. Also find optimal value of TVC.
51. A manufacturer has to supply his customer with 24,000 units of his product per year. This demand is fixed and known. Since the unit used by the customer is an assembly-line operation and the customer has no storage space for the units, the manufacturer must ship a day's supply each day. If the manufacturer fails to supply the required units, he will lose the account and probably his business. Hence, thecost of shortage is assumed to be infinite, and, consequently, none will be tolerated. The inventory holding cost amount to 0.10 per unit per month, and set-up cost per run is Rs. 350. Find the optimum lot size and the length of optimum production run.
52. For an item, the production is instantaneous. The cost of one item is Re 1 per month and the set-up cost is Rs. 25. If the demand is 200

units per month, find the optimum quantity to be produced-per set-up and hence determine the total cost of storage and set-up per month.

53. 'Although the classic inventory model known as EOQ model is too over simplified to represent many of the real world situations, it is an excellent starting point from which to develop more realistic and complex inventory decision models.' In the light of the statement, explain the major limitations of the EOQ model and how these are sought to be overcome.

54. In deterministic lot size models, what additional cost factor must be considered when price breaks are involved, as compared to those models with no price breaks? Explain.

55. A manufacturer has to supply his customer with 600 units of his product per year. Shortages are not allowed and the storage cost amounts to Rs. 0.60 per unit per year. The set-up cost per run is Rs. 80.00. Find the optimum run size and the minimum average yearly cost.

56. A certain item costs Rs. 235 per tonne. The monthly requirement is 5 tonnes and each time the stock is replenished there is a set-up cost of Rs. 1,000. The cost of carrying inventory has been estimated at 10 per cent of the value of the stock per year. What is the optimal order quantity?

57. In a paints manufacturing unit, each type of paint is to be ground to a specified degree of fineness. The manufacturer uses the same mill for a variety of paints and after completion of each batch, the mill has to be cleaned and the ball charge properly made up. The set-up cost from one type of paint to another is estimated to Rs. 80 per batch. The annual sales of a particular grade of paint are 30,000 litres and the inventory carrying cost is Re 1 per litre. Given that the rate of production is 3 times the sales rate, determine the economic batch size.

58. Find the most economic batch quantity of a product on a machine if the production rate of that item on the machine is 200 pieces/day and the demand is uniform at the rate of 100 pieces/day. The set-up cost is Rs. 200 per batch and the cost of holding one item in inventory is Re 0.81 per day. How will the batch quantity vary if the machine production rate was infinite?

59. An aircraft company uses rivets at a constant rate of 2,500 per year. Each unit costs Rs. 30 and the company personnel estimate that it

costs Rs. 130 to place in order, and that the carrying cost of inventory is 10 per cent per year. How frequently should orders be placed? Also detemine the optimum size of each order.

60. If in Model I(a), the set-up cost instead of being fixed is equal to $(C_p + B_q)$, where B is the set-up cost per unit item produced, then show that there is no change in the optimum order quantity produced due to this change in the set-up cost.

61. Yogesh keeps his inventory in special containers. Each container occupies 10 sq ft of store space. Only 5,000 sq ft of the 9,000 containers, priced at Rs. 8 per container. The ordering cost is estimated at Rs. 40 per order, and the annual carrying costs amount to 25 per cent of the inventory value. Would you recommend to Yogesh to increase his storage space? If so, how much should be the increase?

62. Explain clearly with suitable examples the different costs that are involved in the inventory problems.

2

Properties of Transportation Problem

INTRODUCTION

The transportation problem is one of the sub-classes of the L.P.S. in which the objective is to transport various amounts of a single homogenous commodity, that one initially stored at various origin, to different destinations in such a way that the total transportation cost is a minimum.

Let us illustrate by formulating a typical transportation problem involving m origins and n destinations.

The transportation problem can be described as follows.

Suppose that the factories F_i $\{i = 1, 2, ..., m\}$ called the origins or sources produce the non-negative quantities $a_i = \{i = 1, 2, - m\}$ of product and non-negative quantities b_j $(j = 1, 2, ..., n)$ of the same product are required at other n places, called the destinations, such that the total quantity produced is equal to the total quantity required

i.e., $$\sum_{i=1}^{m} a_i = \sum_{j=1}^{n} bj \qquad ...(1)$$

Also support that C_{ij} is the cost of transportation of on unit from the ith source to the jth destination. Then the problem is to determine x_{ij}, the quantity transported from the ith source to the jth destination,

MATHEMATICAL FORMULATION OF TRANSPORTATION PROBLEM

Let there be m origins and n destinations (n may or may not be equal to m) with ith origin possessing ai units of a certain product and jth destination requiring b_j units of the same product.

Assume that the total available is equal to the total required

i.e., $$\sum_{i=1}^{m} a_i = \sum_{j=1}^{n} b_j \qquad ...(1)$$

Let c_{ij} be the cost of transportation of one unit product from ith origin to jth destination and x_{ij} be the quantity transported from the ith origin to the jth destination. Then the problem is to determine non-negative (≥ 0) values of xij, satisfying the availability restrictions as well as the requirement restrictions, in such a way that the total transportation cost is minimized.

i.e., find x_{ij} (≥ 0) for i = 1, 2, ... , m; j = 1, 2, ... , n which minimize

$$Z = \sum_{i=1}^{m}\sum_{j=1}^{n} c_{ij}x_{ij} \qquad ...(2)$$

such that $$\sum_{j=1}^{n} x_{ij} = a_i,\ i = 1, 2, ... ,m \qquad ...(3)$$

and $$\sum_{i=1}^{m} x_{ij} = b_j,\ j = 1, 2, ... ,n \qquad ...(4)$$

The equations (3) and (4) may be called the row and column equations respectively.

In such a way that the total transportation cost $\sum_{i=1}^{m}\sum_{J=1}^{n} c_{iJ}x_{iJ}$ is minimized.

The transportation problem as described above can be represented in a tabular form as follows:

Destination→ Sources↓	**w_1**	**w_2**	**w_j**	**w_n**	**Capacities of the sources**
F_1	c_{11}	c_{12}	c_{1J}	c_{1n}	a_1
F_2	c_{21}	c_{22}	c_{2J}	c_{2n}	a_2
F_i	c_{i1}	c_{i2}	c_{ij}	c_{in}	a_i
F_m	c_{m1}	c_{m2}	c_{mJ}	c_{mn}	a_m
Requirements→	b_1	b_2	b_J	b_n	$\sum_{i=1}^{m} a_i \sum_{J=1}^{n} b_J$.

The calculations are directly on the transportation array given below which gives the curred trial solution.

Destination→ Sources↓	w_1	w_2	w_J	w_n	Capacities of the sources
F_1	x_{11}	x_{12}	x_{1j}	w_{1n}	a_1
F_2	x_{21}	x_{22}	x_{2j}	w_{2n}	a_2
F_i	x_{ij}	x_{i2}	x_{iJ}	x_{in}	a_i
F_m	x_{m1}	x_{m2}	x_{mJ}	x_{mn}	a_m
Requirements→	b_1	b_2	b_J	b_n	$\sum_{i=1}^{m} a_i \sum_{J=1}^{n} b_J$.

The above two tables can be combned together by writing the asts CiJ within the brocket ( ), as follow:

Destination→ Sources↓	w_1	w_2	w_J	w_n	Capacities of the sources
F_1	$x_{11}(c_{11})$	$x_{12}(c_{12})$	$x_{1J}(c_{1J})$	$w_{1n}(c_{1n})$	a_1
F_2	$x_{21}(c_{21})$	$x_{22}(c_{22})$	$x_{2J}(c_{2J})$	$w_{2n}(c_{2n})$	a_2
F_i	$x_{ij}(c_{12})$	$x_{i2}(c_{12})$	$x_{iJ}(c_{iJ})$	$x_{in}(c_{in})$	a_i
F_m	$x_{m1}(c_{m1})$	$x_{m2}(c_{m2})$	$x_{mJ}(c_{mJ})$	$x_{mn}(c_{mn})$	a_m
Requirements→	b_1	b_2	b_J	b_n	$\sum_{i=1}^{m} a_i \sum_{J=1}^{n} b_J$.

RELATION BETWEEN AN ASSIGNMENT PROBLEM AND TRANSPORTATION PROBLEM

An assignment problem is a special case of the transportation problem when each origin is associated with one and only one destination. The numerical evaluations of such association are called 'effectiveness' in place of 'transportation costs'. In this case m = n, all a_i and b_j are unity and each x_{ij} is limited to one of the two values 0 and 1. In these circumstances exactly of x_{ij} can be non-zero (*i.e.*, 1), one for each origin and one for each destination.

FEASIBLE SOLUTION, BASIC FEASIBLE SOLUTION AND OPTIMUM SOLUTION OF TRANSPORTATION PROBLEM

Now we shall define few terms that are used in the transportation problem.

(i) **Feasible Solution (F.S.) :** A feasible solution to a transportation problem is a sets 0 non-negative individual allocations ($x_{ij} \geq 0$) which satisfies the row and column sum restriction [*i.e.*, equations (3) and (4) of 2].

(ii) **Basic Feasible Solution (B.F.S.) :** A feasible solution of an m by n transportation problem is said to be basic if the total number of positive allocations is equal to $m + n - 1$ *i.e.*, one less than the sum of the number of rows and columns.

(iii) **Optimum solution.** A feasible solution (not necessarily basic) is said to be optimum if it minimizes the total transportation cost.

EXISTENCE OF REUSABLE SOLUTION

Theorem:

A necessary and sufficient condition for the existence of feasible solution of a transportation problem is

$$\Sigma a_i = \Sigma b_j \quad (i = 1, 2, \ldots, m;\ j = 1, 2, \ldots, n).$$

Proof:

The condition is necessary: Let there exist a feasible solution to the transportation problem. Then

$$\sum_{j=1}^{n} x_{ij} = a_i,\ i = 1, 2, \ldots, m$$

and
$$\sum_{i=1}^{m} x_{ij}\ j = b_j = 1, 2, \ldots, n.$$

Summing over all i and j respectively, we get

$$\sum_{i=1}^{m}\sum_{j=1}^{n} x_{ij} = \sum_{i=1}^{m} a_i \quad \text{and} \quad \sum_{j=1}^{n}\sum_{i=1}^{m} x_{ij} = \sum_{j=1}^{n} b_j$$

$$\Rightarrow \quad \sum_{i=1}^{m} a_i = \sum_{j=1}^{n} b_j.$$

The condition is sufficient: Let $\sum_{i=1}^{m} a_i = \sum_{j=1}^{n} b_j = k$ (say).

If $x_{ij} = \lambda_i b_j$ for all i and j, where $\lambda_i \neq 0$ is any real number, then

$$\sum_{j=1}^{n} x_{ij} = \sum_{j=1}^{n} \lambda_i b_j = \lambda_i \sum_{j=1}^{n} b_j = k\lambda_i$$

$$\Rightarrow \quad \lambda_i = \frac{1}{k}\sum_{j=1}^{n} x_{ij} = \frac{a_i}{k}.$$

Thus, $x_{ij} = \lambda_i b_j = \frac{a_i b_j}{k}$, for all i and j.

As $a_i \geq 0$, $b_j \geq 0$ so $x_{ij} \geq 0$ for all i and j.

Hence a feasible solution exists.

Note: Such a transportation problem in which $\sum a_i = \sum b_j$ is termed as balanced transportation problem. Hence a balanced transportation problem always has a F.S.

Difference between a Transportation and on Assignment Problem

An assignment problem is a special case of the transportation problem is which m = n, all the a_i and b_j are unity, and each x_{ij} is limited to one of the two value 0 and 1.

In these circumstances, exactly n of x_{ij} can be non-zero, one in each row of the array and one in each column showing that only one source (person) can be assigned to each destination (job).

Some important definitions.

1. *A feasible solution:* A feasible solution to a transportation problem is a set of non-negative individual all cations ($x_{ij} \geq 0$) which satisfies the row and column sum restrictions.
2. *Basic feasible solution:* A feasible solution of a m by n transportation problem is said to be a basic feasible solution if the total number of positive allocations x_{iJ} is exactly equal to m + n – 1, *i.e.* one less than the sum of the number of rows and columns.
3. *Optimal solution:* A feasible solution (necessarily basic) is said to be optimal it minimizes the total transportation cost.
4. *Now-degenerate basic feasible solution:* A feasible solution of m by n transportation problem is said to be non degenerate basic feasible solution if.

(i) Total number of positive allocations is exactly equal to (m + n – 1).

(ii) these allocations are in independent portions.

BASIC FEASIBLE SOLUTION OF A TRANSPORTATION PROBLEM

A transportation problem is a special case of a linear programming problem. So the definition of B.F.S. is same as given earlier for L.P.P. But we find that in a transportation problem out of an unknowns there are only m + n – 1 basic variables. This happens due to redundancy in the constraints of the transportation problem. The condition $\sum a_i = \sum b_j$ can be used to reduce one constraint. This can be easily justified by proving the following theorem:

Theorem:

Out of (m + n) equations, there are only m + n – 1 independent equations in a transportation problem, m and n being the number of origins and destinations and any one equation can be dropped as the redundant equation.

Proof:

Consider m row equations and n – 1 column equations of the transportation problem as

$$\sum_{j=1}^{n} x_{ij} = a_i,\ i = 1, 2, \ldots, m \qquad \text{...(1)}$$

and $$\sum_{i=1}^{m} x_{ij} = b_j,\ j = 1, 2, \ldots, n-1 \qquad \text{...(2)}$$

Now adding m origin constraints given in (1), we get

$$\sum_{i=1}^{m}\sum_{j=1}^{n} x_{ij} = \sum_{i=1}^{m} a_i \qquad \text{...(3)}$$

Also, adding (n – 1) destination constraints given in (2), we get

$$\sum_{j=1}^{n-1}\sum_{i=1}^{m} x_{ij} = \sum_{j=1}^{n-1} b_j \qquad \text{...(4)}$$

Subtracting (4) from (3), we get

$$\sum_{i=1}^{m}\sum_{j=1}^{n} x_{ij} - \sum_{j=1}^{n-1}\sum_{i=1}^{m} x_{ij} = \sum_{i=1}^{m} a_i - \sum_{j=1}^{n-1} b_j$$

or $$\sum_{i=1}^{m}\left(\sum_{j=1}^{n} x_{ij} - \sum_{j=1}^{n-1} x_{ij}\right) = \sum_{j=1}^{n} b_j - \sum_{j=1}^{n-1} b_j \qquad [\because \sum a_i = \sum b_j]$$

or $$\sum_{i=1}^{m} x_{in} = bn,$$ which is the nth destination-constraint.

It follows that if m + n – 1 constraints are satisfied then the (m + n)th constraint will be automatically satisfied due to the condition $\sum a_i = \sum b_j$. Thus we have only (m + n – 1) linearly independent equations. Out of (m + n) equations, one (any) is redundant.

Hence the theorem is proved.

It indicates that B.F.S. will contain almost $m + n - 1$ positive variables, others being zero.

EXISTENCE OF AN OPTIMAL SOLUTION

Theorem:

There always exists an optimal solution to balanced transportation problem.

Proof:

We have $\sum_{i=1}^{m} a_i = \sum_{j=1}^{n} b_j$.

It follows that a feasible solution exists of the problem *i.e.*, $x_{ij} \geq 0$ for all i and j.

Form the constraints of the problem each $x_{ij} \leq \min(a_i, b_j)$.

Thus $0 \leq x_{ij} < \min(a_i, b_j)$ *i.e.*, the feasible region of the problem is non-empty, closed and bounded.

Hence, there exists an optimal solution.

LOOPS IN TRANSPORTATION TABLE AND THEIR PROPERTIES

Loop

Definition : *An ordered set of four or more cells is said to form a loop if it has the following properties:*

(i) *Any two adjacent cells of the set lie either in the same row or in the same column and*

(ii) *no three or more adjacent cells lie in the same row or in the same column.*

The first cell of the set will follow the last one in the set.

We get a closed path satisfying the above conditions (i) and (ii) if we join the cells of loop by horizontal and vertical line segments.

Consider two sets L = {(1, 1), (4, 1), (4, 4), (2, 4), (2, 3), (1, 3)}

and L' = {(3, 1), (3, 4), (2, 4), (2, 3), (2, 2), (1, 2), (1, 1)}

where (i, j)th cell of the transportation table is denoted by (i, j). Then it can be observed that the set L forms a loop while the set L' does not form a loop, because the three cells (2, 4), (2, 3) and (2, 2) lie in the same row. Diagrammatic illustration is given below:

Loop

(1, 1)	(1, 3)		
		(2, 3)	(2, 4)
(4, 1)			(4, 4)

Non-Loop

(1, 1)	(1, 2)		
	(2, 2)	(2, 3)	(2, 4)
	(3, 1)		(3, 4)

Properties

(i) Every loop has an even number of cells.

(ii) a feasible solution to a transportation problem is basic if and only if, the corresponding cells in the transportation table do not form a loop.

SOLUTION OF A TRANSPORTATION PROBLEM

The solution of a transportation problem consists of the following two steps:

Step 1. To find an initial basic feasible solution.

Step 2. To obtain an optimal solution by making successive improvements to initial basic feasible solution (obtained in step 1) until no further decrease in the transportation cost is possible.

BASIC FEASIBLE SOLUTION

Here we describe some simple methods to obtain the initial basic feasible solution.

Method 1: North-West Corner Rule

In this rule we have the following steps:

Step 1. Start with the cell (1, 1) at the north-west corner *i.e.*, the top-most left corner and allocate there maximum possible amount. Thus $x_{11} = \min(a_1, b_1)$.

Step 2.

(i) If $b_1 < a_1$ then $x_{11} = b_1$ and there is still some quantity available left in row 1. So move to the right hand cell (1, 2) and make the second allocation of amount $x_{12} = \min(a_1 - x_{11}, b_2)$ in the cell (1, 2).

(ii) If $b_1 > a_1$, then $x_{11} = a_1$ and there still some requirement left in column 1. So move vertically downwards to the cell (2, 1) and make the second allocation of amount $x_{21} = \min(a_1, b_1 - x_{11})$ in this cell.

(iii) If $b_1 = a_1$ then $x_{12} = 0$ or $x_{21} = 0$.

Start from the new north-west corner of the transportation table and allocate there as much as possible.

Step 3. Repeat steps 1 and 2 until all the available quantity is exhausted or all the requirement is satisfied.

The following example explains the method:

Example:

Find the initial basic feasible solution of the following transportation problem:

		To			
		W_1	W_2	W_3	Available
	F_1	2	7	4	5
	F_2	3	3	1	8
From	F3	5	4	7	7
	F_4	1	6	2	14
Requirement		7	9	18	34

Solut on:

First we construct an empty 4 by 3 matrix complete with row and column requirements.

Start with the cell (1, 1) at the north-west corner (top-most left corner) and allocate it maximum possible amount. Thus x_{11} = 5 as minimum of a_1 = 5 and b_1 = 7 is 5.

		To			
		W_1	W_2	W_3	Available
	F_1	5 (2)			5
	F_2	2 (3)	6 (3)		8
From	F_3		3 (4)	4 (7)	7
	F_4			14 (2)	14
Requirement		7	9	18	

There is no amount left available at source 1 so in place of moving to right we move vertically downwards to the cell (2, 1) and allocate as much as possible there. The column 1 still needs the amount 2 and the amount 8 is available in row 2 so we allocate the maximum amount 2 to the cell (2, 1) *i.e.*, x_{21} = 2. Thus allocations for column 1 are complete. Now we move to the right of the cell (2, 1). Since the amount 6 is still available in row 2 and amount 9 is needed in column 2, so we allocate the maximum amount 6 is still available in row 2 and amount 9 is needed in column 2, so we allocate the maximum amount 6 in the cell (2, 2) *i.e.*, x_{22} = 6. This completes the allocations for row 2. Now we move vertically downwards to the cell (3, 2). In column 2 the amount 3 is still needed and in row 3 amount available is 7 so we allocate the maximum amount 3 in the cell (3, 2) *i.e.*, x_{32} = 3. Thus allocations for column 2 are complete. Now the amount 4 is still available in row 3 and amount 18 is needed in column 3, so we move to the cell (3, 3) and allocate the maximum amount 4 to this cell *i.e.*, x_{33} = 4. Thus, there is no amount left available at source 3. Now we move downwards to the cell (4, 3). The amount 14 is still needed in column 3 and an equal amount 14 is available in row 4, so we allocate the amount 14 to cell (4, 3) *i.e.*, x_{43} = 14. The resulting feasible solution is shown in the above table. Allocations in the cells are in such a way that the total the total in each row and each column is the same as shown against the respective rows and columns Multiplying each individual allocation by its corresponding unit cost in ( ), and adding, the total cost corresponding to this feasible solution is = 5.2 + 2.3 + 6.3 + 3.4 + 4.7 + 14.2 = Rs. 102.

Note:

In this method we always move to the right or down, so no loop can be formulated by drawing horizontal and vertical lines to the allocations. Also at each step (allocation) at least one row or column is discarded from further consideration, while the last allocation discards both a row and a column simultaneously, so we cannot get more than (m + n – 1) individual positive allocations. Thus *we always get a non-degenerate basic feasible solution by the north west corner rule.*

Method 2: Lowest Cost Entry (Matrix Minima) Method

In this method we have the following steps:

Step 1. Examine the cost matrix carefully and find the lowest cost. Let it be c_{ij}. Then allocate x_{ij} as much as possible in the cell (i, j). x_{ij} = min (a_1, b_j).

Step 2.

(i) If $x_{ij} = a_1$, then the capacity of the ith origin is completely exhausted. In this case cross out the ith row of the transportation table and decrease the requirement b_j by a_1. Now go to step 3.

(ii) If $x_{ij} = b_j$, then the requirement of jth destination is completely satisfied. In this case cross out the jth column of the transportation table and decrease a_i by b_j. Now go to step 3.

(iii) If $x_{ij} = a_i = b_j$, then either cross-out the ith row or jth column but not both. Now go to step 3.

Step 3. Repeat steps 1 and 2 for the reduced transportation table until all the available is exhausted or all the requirement is satisfied.

Note:

If the cell of lowest cost is not unique. We can select any one of these cells. The method is well explained by taking the same numerical example as in method 1.

		To			
		W_1	W_2	W_3	
	F_1	(2)	2(7)	3(4)	5
From	F_2	(3)	(3)	8(1)	8
	F_3	(5)	7(4)	(7)	7
	F_4	7(1)	(6)	7(2)	14
		7	9	18	

First we write the cost and requirement matrix. We examine the cost matrix and find that there is lowest cost 1 in cell (2, 3) and in (4, 1). We choose any one of these. Say the cell (2, 3) and allocate the maximum possible amount 8 to this cell. This exhausts the availability from F_2. Now leaving the second row in the reduced transportation table we find that there is lowest cost 1 in cell (4, 1). Here we allocate the maximum possible amount 7. This satisfies the requirement of W_1.

Leaving the first column, in the reduced transportation table we find the lowest cost 2 in the cell (4, 3). We allocate the maximum possible amount 7 to this cell. This exhausts the availability from F_4. Leaving the 4th row, in the reduced transportation table, we find that there is lowest cost 4 in cell (1, 3) and in cell (3, 2). We allocate the amount 3 in the cell (1, 3). This satisfies the requirement of W_3. In order to satisfy the availability of F_1 we allocate the amount 2 to the cell (1, 2). To complete the requirement of 9 units in column 2, we allocate the amount 7 to the cell (3, 2). Thus we get the require B.F.S. shown in the above table.

The transportation cost = 2.7 + 3.4 + 8.1 + 7.4 + 7.1 + 72 = Rs. 83.

Method 3: Unit Cost Penalty Method (Vogel's Approximation Method).

In this method we have the following steps:

Step 1. Identify the smallest and next to smallest costs for each row of the transportation table. Find the difference between them for each row. Write these differences alongside the transportation table against the respective rows by enclosing them in parentheses. Write the similar differences for each column below the corresponding column. These are called 'penalties'.

Step 2. Now select the row or column for which the penalty is the largest. If a tie occurs, use any arbitrary tie breaking choice. Allocate the maximum possible amount to the cell with lowest cost in that particular row or column.

Let the largest penalty correspond to ith row and let c_{ij} be the smallest cost in the ith row. Allocate the amount x_{ij} = min (a_i, b_j) in the cell (i, j).

Then we cross out the ith row or the jth column in the usual manner and construct the reduced matrix.

Step 3. Now compute the row and column penalties for the reduced transportation table and repeat the step 2. We continue this process until all the available quantity is exhausted or all the requirement is satisfied.

The method is well explained by taking the same numerical example as in method 1.

First we write the cost and requirement matrix and compute the penalties as follows:

	W_1	W_2	W_3	Available	Penalties
F_1	5(2)	(7)	(4)	5	(2)
F_2	(3)	(3)	(1)	8	(2)
F_3	(5)	(4)	(7)	7	(1)
F_4	(1)	(6)	(2)	14	(1)
Requirement	7	9	18		
Penalties	(1)	(1)	(1)		

We find that the maximum penalty (2) is associated with row 1 and row 2, so we may select any one of these. If we select row 1, then we allocate the maximum possible amount to the lowest cost cell in this row *i.e.*, cell (1, 1). Thus, x_{11} = min (5, 7) = 5. This exhausts the avaiiability from F_1. So we cross the row 1. Leaving this row, the reduced cost and requirement matrix is as follows:

	W_1	W_2	W_3	Available	Penalties
F_2	(3)	(3)	8(1)	8	(2)
F_3	(5)	(4)	(7)	7	(1)
F_4	(1)	(6)	(2)	14	(1)
Requirement	2	9	18		
Penalties	(2)	(1)	(1)		

We note that the amount still needed to column 1 is 2.

Since the maximum penalty (2) is associated with row 1 and column 1 so we may select any one of these.

We select row 1 of this table and allocate the maximum possible amount 8 to the cell with cost 1 (lowest) in this row. This exhausts the availability from F_2 and leaves the requirement 10 for W_3. Again leaving row of F_2, the reduced transportation table is as follows:

	W_1	W_2	W_3	Available	Penalties
F_3	(5)	(4)	(7)	7	(1)
F_4	(1)	(6)	10(2)	14	(1)
Requirement	2	9	10		
Penalties	(4)	(2)	(5)		

In this table, the maximum penalty (5) is associated with column 3 so the maximum possible amount 10 is allocated to the cell with lowest cost 2 in this column. This completes the requirement of W_3. After leaving the column corresponding to W_3 the remaining table is as follows:

	W_1	W_2	Available	Penalties
F_3	(5)	7(4)	7	(1)
F_4	2(1)	2(6)	4	(5)
Requirement	2	9		
Penalties	(4)	(2)		

In this table, the maximum penalty (5) is associated with row 2, so the maximum possible amount 2 is allocated to the cell with lowest cost 1 in this row. The remaining amount 2 available to F_4 is allocated to the cell with cost 6. In the last to meet the requirement of W_2 the amount 7 is allocated to the cell with cost 4.

Thus we get the required B. F. S. as shown in the table.

	W_1	W_2	W_3	Available
F_1	5(2)	(7)	(4)	5
F_2	(3)	(3)	8(1)	8
F_3	(5)	7(4)	(7)	7
F_4	2(1)	2(6)	10(2)	14
Requirement	7	9	18	

The total transportation cost

$$= 5.2 + 8.1 + 7.4 + 2.1 + 2.6 + 10.\ 2 = \text{Rs. } 80.$$

Note : If in the selected row or column the minimum cost is cost not unique, then allocate in that cell in which more allocations can be made.

Although, Vogel's method takes more time as compared to other methods, but it reduces the time in reaching the optimal solution. To obtain the optimal solution the students are advised to find the initial B. F. S. by Vogel's method.

SOLUTION OF A TRANSPORTATION PROBLEM

The solution (optimal) of a transportation problem consists of the following two steps.

Step 1. To find on initial basic feasible solution.

Step 2. To obtain an optimal solution making improvements to initial basic passible solution until no further decrease in the transportation co: is possible.

To find an initial feasible solution.

There are several methods for finding the initial feasible solution of the given transportations problem. Here we describe the following three methods

Method 1: North-West Corner Rule

By this rule we allocate a set of allocations in the cells so that the row totals and columns totals will be as indicted before each, as follows.

(i) Start with the cell (1, 1) at the north-west corner *i.e.* the top most left corner and allocate it maximum possible amount.

(ii) Then move to the right hand cell (1, 2) if there is sill any available quantity left, dowries move to the down cell (2, 1) and allocate it maximum possible amount.

(iii) Repeat the step (ii) again and continue until all the available quantity is exhausted.

The method is well explained by toping the following numerical example:

To

		W_1	W_2	W_3	Supply
	F_1	(2)	(7)	(9)	5
From	F_2	(3)	(3)	(1)	8
	F_3	(5)	(4)	(7)	7
	F_4	(1)	(6)	(2)	14
Demand		7	9	18	34

To

		W_1	W_2		
	F_1	5 (2)			$5 = a_1$
From	F_2	2 (3)	6 (3)		$8 = a_2$
	F_3		3 (4)	4 (7)	$7 = a_3$
	F_4			14 (2)	$14 = a_4$
		7 b_1	$9 = b_2$	$18 = b_3$	

(i) We start with the topmost left corner and allocate is maximum possible amount 5. (Since mini of $a_1 = 5$ and $b_1 = 7$).

(ii) Since there is no amount left available at source 1 so we move downwards to the cell (2, 1), in place of moving to right, and allocate it maximum possible amount. Since the column 1 still need the amount 2 and the amount 8 is available in row 2 so we allocate the maximum amount 2 to this cell (2, 1). Now the allocation for column 1 is complete, so we move to the right of this cell. Since the amount 6 is still available in row 2 and amount 9 is needed in column 2, so we allocate the maximum amount 6 in this cell (2, Thus, allocation for row 2 is complete, so we move downwards to the cell (3, 2). Since the amount 3 is still needed in column 2 and amount 7 is available in row 3, so we allocate the maximum amount 3 in this cell (3, 2). Thus allocation for column is complete, so we move to the right of this cell. Since the amount 4 is still available in row 3 and amount 19 is needed in column 3, so we allocate the maximum amount 4 to the cell (3, 3). Thus the requirements of the row 3 is complete, so we more to the downgrade cell (4, 3). Since amount 14 is still available in row 3 and an equal amount 14 is needed in row 4, so we allocate this amount 14 to this cell (4, 3). This is complete the allocation.

In the end it may be checked that the sum of rows and columns are as needed. On multiplying each individual allocation by its corresponding unit in ( ), and adding, the total transportation cost do this F. S is

$$= 5 \times 2 + 2 \times 3 + 6 \times 3 + 3 \times 4 + 4 \times 7 + 14 \times 2$$

$$= \text{Rs. } 102.$$

In this north-west corner rule we more to the right or down, so no loop can be formulated here by drawing horizontal and vertical lines to the allocation. Also at each step at least one row or column is discarted from further consideration and at the last allocation both rows and columns are discarded. So we cannot get more than (m + n – 1) individual positive allocation by this rule. Thus we always get a non degenerate basic feasible solution by this north-west corner rule.

Method 2: Lowest Cost Entry Method (Layman or Method of Matrix Minima)

In this we write the costs and the requirement matrix. The cost are written within brackets ( ). Now we examine the cost matrix carefully and choose the cell with lowest cost and allocate there as much as passible. If such cell of lowest cost in not unique, we can select any one of these cells. Again we examine the cost matrix and select the cell with the lowest cost (the cell in which allocation has been made is not considered) and allocate there as much as possible we continue this process unit all the available quantify is exhausted.

The method is well explained by taping the same number example in method 1. First we write the cost and requirements matrix as following.

		To			
		W_1	W_2	W_3	
	F_1	(2)	2(7)	3(4)	5
From	F_2	(3)	(3)	8(1)	8
	F_3	(5)	7(4)	(7)	7
	F_4	7(1)	(6)	7(2)	14
		7	9	18	

Examine the cost matrix in above table we find there is lowest cost 1 in cell (2, 3) and in (4, 1) we choose any one of these say the cell (2, 3) and allocate the maximum amount 8 to this cell Leaving this cell we find that there is lowest best 1 in cell (4, 1) where we allocate the maximum amount 7. Continuing in this way we get the required feasible solution shows in always table.

The total transportation cost to this F. S. is

$$= 2 \times 7 + 3 \times 4 + 8 + 1 + 7 \times 4 + 7 \times 1 + 7 \times 2$$

$$= \text{Rs. } 83.$$

This cost is less than the cost associated with the F.S. obtained by north west corner rule.

The initial feasible solution obtained by this method usually gives a lower transportation cost than that obtained is north west corner rule.

Method 3: Unit Cost Penalty Method (or Vogel's Approximation Method)

In this method we write the differences of the smallest and the second smallest costs in each column, below the correspond column and write the similar differences of each row to the right of the corresponding row. These individual different can be thought of penalty for making allocations in the second lowest cost cell instead of lowest cost cell in each row or column. Now we select the row or column for which the penalty is the largest and allocate the maximum possible amount to the cell with lowest cost in that particular row or column. If there are more than one largest penalty rows or column, then we may select any one of them. Then we cross that row or column in which the requirement has been satisfied and construct the reduced matrix. We continue this process on the reduced matrices till all the allocations have been made.

The method is well explain by taking the numerical examples as in method I.

First we write the cost and requirement matrix as follows:

	W_1	W_2	W_3	Available	penalties
F_1	5(2)	(7)	(4)	5	(2)
F_2	(3)	(3)	(1)	8	(2)
F_3	(5)	(4)	(7)	7	(1)
F_4	(1)	(6)	(2)	14	(1)
Demand	7	9	18		
Penalties	(1)	(1)	(1)		

Since the maximum penalty (2) is associated with row 1 and row 2, so we may select any one of these. Suppose we select row 1 then we allocate the maximum possible amount 5 to the cell (1, 1) with lowest cost and cross this row 1.

The first reduced matrix after leaving row with remaining demand and available is and follows.

(Note that the amount 5 has been allocated to the column 1, so the amount still needed to column 1 is 2)

		W_1	W_2	W_3	Available	Penalties
	F_2	(3)	(3)	8(1)	8	(2)
	F_3	(5)	(4)	(7)	7	(1)
	F_4	(1)	(6)	(2)	14	(1)
Demand		2	9	18		
Penalties		(2)	(1)	(1)		

Since the maximum penalty (2) is associated with row 1 above table, so the maximum possible amount 8 is allocated to the cell (1, 3) with lowest cot in this row.

The second reduced matrix after leaving row 1 of the matrix in above table with remaining demands and available is as follows

		W_1	W_2	W_3	Available	Penalties
	F_3	(5)	(4)	(7)	7	(1)
	F_4	(1)	(6)	10(2)	14	(1)
Demand		2	9	10		
Penalties		(4)	(2)	(5).		

Since the maximum penalty (5) is associated with column 5 of matrix in above table, so the maximum possible amount 10 is allocated to the cell (2, 3) with lowest cost in this column.

The third reduced matrix after learning column 3 of the matrix in above table with remaining demands and availables is as follows:

		W_1	W_2	Available	Penalties
	F_3	(5)	7(4)	7	(1)
	F_4	2(1)	2(6)	4	(5)
Demand		2	9		
Penalties		(9)	(2)		

Since the maximum penalty (5) is associated with row 2, so the maximum possible amount 2 is allocated to the cell (2, 1) with lowest cost in this row. The remaning amount 2 in row second is allocated to the cell (2, 2) and then we allocate the renaming amount 7 to the cell (1, 2) with minimum cost after crossing the row 2.

Thus we get the required feasible solution as shown in below table:

	W_1	W_2	W_3	Available
F_1	5(2)	(1)	(4)	5
F_2	(3)	(3)	8(1)	8
F_3	(5)	7(4)	(7)	7
F_4	2(1)	2(6)	10(2)	14
Demand	7	9	18	

The total transportation cost to this F.S.

$= 5 \times 2 + 8 \times 1 + 7 \times 4 + 2 \times 1 + 2 \times 6 + 10 \times 2$

= Rs. 80.

Which is less than the cost associated with the feasible solution obtained by the previous method 1 and 2.

Theorem:

It we have a B.F.S consisting of $(m + n - 1)$ independent positive allocations, and a set of arbitrary numbers u_i and v_j, $i = 1, 2, \ldots, m$,

$j = 1, 2, n$, such that

$$c_{rs} = u_r + v_s$$

For all occupied cells (r, s), then the evaluation d_{ij} corresponding to each empty cell (i, j) is given by

$$d_{ij} = c_{ij} - (u_i + v_j).$$

Proof:

Mathematically a transportation problem is to determine $x_{ij} \geq 0$, which minimize.

$$Z = \sum_{i=1}^{m} \sum_{J=1}^{n} c_{iJ} \cdot x_{iJ} \qquad \ldots(1)$$

subject to the restrictions

$$\sum_{J=1}^{n} x_{ij} = a_1 \text{ or } 0 = a_i - \sum_{J=1}^{n} x_{ij} \qquad \ldots(2)$$

for $i = 1, 2, \dots, m$

$$\sum_{i=1}^{m} x_{iJ} = b_j \text{ or } 0 = b_j - \sum_{i=1}^{m} x_{iJ} \quad ...(3)$$

for $j = 1, 2, \dots, n$.

Multiplying (2) by u_i ($i = 1, 2..., m$), 3 by v_j (= 1, 2...,n) and adding to (11) we have

$$Z = \sum_{i=1}^{m}\sum_{J=1}^{n} c_{iJ} x_{iJ} + \sum_{i=1}^{m} u_i \left(a_i - \sum_{J=1}^{n} x_{iJ} \right) + \sum_{J=1}^{n} v_J \left(b_j - \sum_{i=1}^{m} x_{iJ} \right)$$

$$\text{or} \quad Z = \sum_{i=1}^{m}\sum_{j=1}^{n} [a_{ij} - (v_i + v_j)]x_{ij} + \sum_{i=1}^{n} v_i a_i + \sum_{j=1}^{n} v_j b_j \quad ...(4)$$

But it is given that for all occupied cells (cells with positive allocations)

$$c_{rs} = u_r + v_s \quad ...(5)$$

∴ In the objective function (4), all the terms of positive allocation vanish as this coefficients are zero. For thus F.S the value of objective function (4), reduce to

$$Z = \sum_{i=1}^{m} u_i a_i + \sum_{j=1}^{n} v_j b_j \quad ...(6)$$

To prove the required result, let us first determine the cell evaluation for the empty cells (i_1 j) when we allocate + 1 unit to this empty cell the number of positive allocations become (m + n), hence they become independent in position and so a closed loop can be formed. Let us closed loop formed by Joining this cell (i, j) to the occupied cells be as shown in fig (1).

Here cell (i, s), r, s) and (r, j) ore occupied cells

∴ $c_{is} = u_i + v_s$, $c_{rs} = u_r + v_s$, $c_{rj} = u_r + v_j$ when we allocate + 1 unit to the empty cell (i, j), then to maintain the row and column sums unchanged we shall decrease the individual allocation at the cell (i, s) by 1, increase that at the less (r, s) by 1 and decrease (i, j) that at cell (r, T) by 1. That the c_{ij} allocation at the occupied cells (i, s), (r, s), (r, T) are changed. Thus, the cell evaluation d_{ij} corresponding to this empty cell (i, j) is given by d_{ij} = Difference in the total cost for the new solution and the original one

$$= c_{iT} - c_{is} + c_{rs} - c_{rj}$$

$$= c_{ij} - (u_i + v_s) + (u_r + v_s) \ (u_r + v_j)$$

$$\text{or} \quad d_{ij} = c_{ij} - (u_i + v_j)$$

This gives the evaluations d_{ij} connecting empty cell (i, j) to the occupied cells by a square shaped loop. By similar reasoning, we way generalize it to the occupied cells.

Thus, we conclude that the evaluation d_{ij} corresponding to each empty cell (i, j) is given by

$$d_{ij} = c_{ij} - (u_i + v_T)$$

This completes the proof of the theorem.

COMPUTATIONAL PROCEDURE OF OPTIMAL TEST

After getting the initial B.F.S of a transportation problem, we test this solution for optometry as follows.

1. For a B.F.S. we determine a set of (m + n) numbers

$$u_i \neq i = 1, 2, \ldots ,m$$

$$v_j \neq J = 1, 2, \ldots ,n$$

Such that for each occupied cell (r, s)

$$c_{rs} = u_r + v_s$$

For this we assign an arbitrary value to one of the u_i's or v_j's then the rest (m + n − 1) of them can early resolved algebraically from the relation $c_{rs} = u_r + v_s$, for occupied cells.

Generally we chooses that u_i or $v_j = 0$ for which the corresponding row or column have the maximum number of undivided allocations.

2. Then we calculate the cell evaluations d_{ij} for each unoccupied cell (i, j) by wring the formula $d_{ij} = c_{ij} - (u_i + v_j)$.
3. Then we examine the matrix of cell evaluations and crackled that
   (i) If all $d_{ij} > 0$, the solution under test is optimal and unique.
   (ii) If all $d_{ij} \geq 0$, with at cost one $d_{ij} = 0$, then the solution under test is optimal and alternative optimal solution exists.
   (iii) If at test one $d_{ij} < 0$, then the solution is not optimal. In the last case we proceed to the next step 4.
4. If at least one $d_{ij} < 0$, in step 3, then we form a new B.F.S. In the next B.F.S we give maximum allocation to the cell for which d_{ij} is minimum and negative, by making an occupied cell empty.
5. Then we repeat the steps 1 to 3 to test the optimality to this new B.F.S.

TO TEST THE SOLUTION FOR OPTIMALITY

Finding the set of u_i (i = 1, 2, 3), v_j (j = 1, 2, 3) such that for occupied cells $c_{rs} = u_r + v_s$ and then entering evaluation $(u_i + v_j)$ and d_{ij} in the unoccupied cells, we get the following table:

(6)		(8)		(4)		u_i
	1		10		3	0
(4)	5	(9)	7	(3)		
					12	–1
	–1		2			
(1)		(2)	3	(6)	–1	
	5					–5
			–1		7	
$v_j \rightarrow$	6		8		4	

Since all d_{ij} are not ≥ 0, the solution test is not optimal.

First Iteration

Here two cell evaluations are negative and they are both most negative. We shall include any one of then in the solution. Let us allocate at cell (2, 1) as much as possible.

Improved solution

			(6)		(8)		(4)	
1–θ	10	3+θ				10		4
			(4)		(9)		(3)	
θ		12–θ		1				11
			(1)		(2)		(6)	
5				5				

Here min $[1 - \theta, 12 - \theta] = 0 \Rightarrow 1 - \theta = 0 \Rightarrow \theta = 1$.

Thus, cell (2, 1) enters the solution while cell (1, 1) leaves the solution *i.e.*, it becomes empty. Transportation cost = 8.10 + 4.4 + 4.1 + 3.11 + 1.5 = Rs. 138.

Second Iteration

Since the largest negative cell evaluation is $d_{32} = -2$, so allocate as much as possible to cell (3, 2)

Next Improved solution

			(6)		(8)		(4)	
	10–θ	4+θ				5		9
			(4)		(9)		(3)	
1+θ		11–θ		6				6
			(1)		(2)		(6)	
5–θ	θ					5		

Here min [5 – θ, 10 – θ] = 0 5 – θ = 0 ⇒ θ = 5.

Thus, cell (3, 2) enters the solution while cell (3, 1) leaves the solution.

Transportation cost = 8.5 + 4.9 + 4.6 + 3.6 + 2.5 = Rs. 128.

To Test the Next Improved Solution for Optimality

	(6)		5	(8)			(4)			u_i↓
					5			9		1
			1							
	(4)			(9)		7	(3)			
		6						6		0
						2.				
	(1)		–1	(2)		(6)			–2	
					5					–5
			2						8	
v_j →		4			7			3		

Since all d_{ij} for empty cells are > 0, so the solution under test is optimal. Thus the solution of the given problem is x_{12} = 5, x_{13} = 9, x_{21} = 6, x_{23} = 6, x_{32} = 5 and minimum cost = Rs. 128.

RULE TO DETERMINE θ

The value of θ, in general, is obtained by equating to zero the of the allocations containing –θ (not + θ) at the corners of the closed loop. Here min [8 – θ, 2 – θ] = 0 or 2 – θ = 0 or θ = 2 units. Thus, cell (2, 2) enters the solution, while cell (4, 2) leaves the solution *i.e.*, it becomes empty. Thus, improved B.F.S. is obtained. The illustration is shown in the following tables:

				(2)		(7)		(4)		
5			5		5					5
				(3)		(3)		(1)		
	θ	8–θ	8				2		6	8
				(5)		(4)		(7)		
	7		7				7			7
				(1)		(6)		(2)		
2	2–θ	10+θ	14		2				12	14
7	9	18			7		9		18	

Total transportation cost foris B.F.S. = 2.5 + 3.2 + 1.6 + 4.7 + 1.2 + 2.12 = Rs. 76, which is less than that for the initial B.F.S.

Now we shall test this improved B.F.S. for optimality.

Step 7. Proceeding as in steps 2, 3 and 4 we get the following table:

	(2)		(7)	5	(4)	3	u_i↓
		5				1	
				2		1	
	(3)	0	(3)		(1)		
				2		6	–1
		3					
	(5)	1	(4)		(7)	2	
				7			0
		4				5	
	(1)		(6)	4	(2)		
		2				12	0
				2			
v_j →		1		4		2	

Step 8. Since all emptiest evaluations (d_{ij}) > 0, the solution under test is optimal.

Thus the solution of the given problem is x_{11} = 5, x_{22} = 2, x_{23} = 6, x_{32} = 7, x_{41} = 2, x_{43} = 12 and minimum transportation cost = Rs. 76.

DEGENERACY IN TRANSPORTATION PROBLEM

We recall that a B.F.S. to an m-origin and n-destination transportation problem will contain almost (m + n – 1) independent non-zero allocations. If this number is exactly (m + n – 1), the B.F.S. is said to be non-degenerate otherwise it is said to be a degenerate one. Thus degeneracy in transportation problem occurs whenever the number of independent individual allocations is cell than (m + n – 1).

Degeneracy in transportation problem can occur in two ways:

(i) B.F.S. may be degenerate from the initial stage onward.

(ii) It may become degenerate at any intermediate stage, when the selection of one entering cell empties two or more pre-occupied cells simultaneously.

In such cases, to resolve degeneracy, we allocate an extremely small amount (close to zero) to one or more empty cells of the matrix (generally lowest cost cells if possible), so that the total number of occupied (allocated cells become $(m + n - 1)$ at independent positions.

The extremely small quantity usually denoted by Δ (delta) or $\in$ (epsilon) satisfies the following conditions:

1. $\Delta < x_{ij}$ for $x_{ij} > 0$
2. $x_{ij} + \Delta = x_{ij} = x_{ij} - \Delta$, $x_{ij} > 0$
3. $\Delta + 0 = \Delta$.
4. If there are more than one Δ's introduced in the solution, then.

(i) if Δ, Δ' are in the same row, $\Delta < \Delta'$ when Δ is to the left to Δ' and

(ii) if Δ, Δ' are in the same column, $\Delta < \Delta'$ when Δ is above Δ'.

Above rules show that even after introducing Δ, the original solution of the problem is not changed. It is merely a technique to apply the optimality test. As Δ has no physical significance, ultimately it is to be omitted.

Following examples will make the procedure clear:

UNBALANCED TRANSPORTATION PROBLEM

If in a transportation problem, the sum of all available quantities is not equal to the sum all requirements i.e., if $\sum_{i=1}^{m} a_i \neq \sum_{j=1}^{n} b_j$ *then such problem is called an unbalanced transportation problem.*

An unbalanced transportation problem may occur in two different forms:

(i) *Shortage in availability, i.e.,* $\sum a_i < \sum b_j$: To modify this type of unbalanced transportation problem to balanced type we introduce a dummy source row in the transportation table. The unit transportation costs from this dummy source to any destination are all set equal to zero. The availability at this dummy source is assumed to be equal to the difference $\sum b_j - \sum a_i$.

(ii) *Excess of availability, i.e.,* $\sum a_i > \sum b_j$. To modify this type of unbalanced transportation problem to balanced type we introduce a dummy destination column in the transportation table. The unit transportation costs from this dummy destination from any source are all set equal to zero. The requirement at this dummy destination is assumed to be equal to the difference $\sum a_i - \sum b_j$.

SOLVED EXAMPLES

Example 1:

Solve the following problem.

	D_1	D_2	D_3	D_4	Supply
O_1	1	2	1	4	30
O_2	3	3	2	1	50
O_3	4	2	5	9	20
Demand	20	40	30	10	100/100.

Solut on:

Step 1. By 'Lowest Cost Entry' method an initial B.F.S of the given problem is given by the following table 1.

Table 1

	D_1	D_2	D_3	D_4	Supply
	1	2	1	4	
O_1	20		10		30
	3	3	2	1	
O_2		20	20	10	50
	4	2	5	9	
O_3		20			20
Demand	20	40	30	10	

Total transportation cost

$= 20 \times 1 + 10 \times 1 + 20 \times 3 + 20 \times 2 + 10 \times 1 + 20 \times 2$

$= 20 + 10 + 60 + 40 + 10 + 40$

Rs. 180.

Step 2. In step 2 we will find u_i and v_j. For this we choose $u_2 = 0$ (Since row 2 contains max. number of allocations.)

$c_{22} = u_2 + v_2$ $\quad$ $c_{23} = u_2 + v_3$ $\quad$ $c_{21} = u_2 + v_4$

$3 = 0 + v_2$ $\quad$ $2 = 0 + v_3$ $\quad$ $1 = 0 + v_4$

$\Rightarrow v_2 = 3$ $\quad$ $\Rightarrow v_3 = 2$ $\quad$ $\Rightarrow v_4 = 1$

Also

$c_{13} = u_1 + v_3$ $\quad$ $c_{11} = u_1 + v_1$ $\quad$ $c_{32} = u_3 + v_2$

$1 = u_1 + 2$ $\qquad 1 = -1 + v_1 \qquad 2 = u_3 + 3$

4 $\quad u_1 = -1 \qquad \Rightarrow v_1 = 2 \qquad u_3 = -1.$

Step 3. Now we find the cell evaluations $u_i + v_j$ for each unoccupied cell (i, J) and enter at upper left corner of corresponding unoccupied cell.

Step 4. Then we find the cell evaluation $d_{ij} = c_{ij} - (u_i + v_j)$ (i.e., the difference of the upper right corner entry from the upper left corner entry) for each unoccupied cell (i, j) and enter at the lower left corner of the corresponding unoccupied cell.

Thus we get the following table 2.

Table 2

		D_1		D_2		D_3		D_4	Supply	u_i
	1		1 2		2 1		1 0	4		
O_1		20				10			30	–1
	0		0		0		4			
	2		3 3		3 2		2 1	1		
O_2				20		20		10	50	0
	1		0		0		0			
	1		4 2		2 1		5 0	9		
O_3				20					20	–1
	3		0		4		9			
Demand		20		40		30		10		
V_J		2		3		2		1		

Since all $d_{ij} \geq 0$, so the solution is optimal.

The solution of the given problem is

From Demand

O_1	to	D_1	Units	20
O_1	to	D_3	Units	10
O_2	to	D_2	Units	20
O_2	to	D_3	Units	20
O_2	to	D_4	Units	10
O_3	to	D_2	Units	20 Ans.

Example 2:

Solve the following problem.

	A_1	A_2	A_3	*Available*
B_1	2	7	4	5
B_2	3	3	7	8
B_3	5	4	1	7
B_4	1	6	2	14
Required	7	9	18	34/34

Solution:

Step 1. By Vogel's method an initial B.F.S. of the given problem is given by the following :

Table 1

	A_1	A_2	A_3	Available
B_1	5 (2)	(7)	(4)	5
B_2	(3)	8 (3)	(7)	8
B_3	(5)	(4)	7 (1)	7
B_4	2 (1)	1 (6)	11 (2)	14
Required	7	9	18	34/34

Total transportation cost

$= 5 \times 2 + 8 \times 3 + 7 \times 1 + 2 \times 1 + 6 \times 1 + 11 \times 2$

$= 10 + 24 + 7 + 2 + 6 + 22$

= Rs. 71.

Step 2. In step 2 we will find u_i and v_j.

For this we choose $u_4 = 0$ (Since row 4 contains maximum number of allocations)

Since $\quad c_{41} = u_4\ v_1 \qquad (\because c_{rs} = u_r + v_s)$

$1 = 0 + v_1$

$\Rightarrow \quad v_1 = 1$

$$c_{42} = u_4 + v_2 \qquad c_{43} = u_4 + v_3$$
$$6 = 0 + v_2 \qquad 2 = 0 + v_3$$
$$v_2 = 6 \qquad v_3 = 2$$

Also
$$c_{11} = u_1 + v_1 \qquad c_{22} = u_2 + v_2 \qquad c_{33} = u_3 + v_3$$
$$2 = u_1 + 1 \qquad 3 = u_2 + 6 \qquad 1 = u_3 + 2$$
$$u_1 = 1 \qquad u_2 = -3 \qquad u_3 = -1.$$

Step 3. Now we find the cell evaluations $u_i + v_j$ for each unoccupied cell (i, j) and enter at upper left corner of corresponding unoccupied cell.

Step 4. Then we find cell evaluations $d_{ij} = c_{ij} - (u_i + v_j)$ (*i.e.,* the difference of the upper right corner entry from the upper left corner entry) for each unoccupied cell (i, j) and enter at the lower left corner of the corresponding uncollected cell.

Thus we get the following table.

Table 2

		A_1			A_2			A_3		Available	ui
	2		2	7		7	3		4		
B_1		5								5	1
	0			0			1				
	–2		3	3		3	–1		7		
B_2					8					8	–3
	5			0			8				
	0		5	5		4	1		1		
B_3								7		7	–1
	5			–1			0				
	1		1	6		6	2		2		
B_4					1			11		14	0
	0			0			0				
Required			7			9			18	34/84	
V_i		1			6			2			

Step 5. Since $d_{32} = -1 < 0$, so the solution is not optimal.

Step 6. Since mini d_{ij} is $d_{32} = -1$ (negative), so we given maximum allocation to this cell from an occupied cell and make necessary changes in other allocations as shown in table 3.

Step 7. The new B.F.S is shown table 3. For this B.F.S total transportation cost

= 5 × 2 + 8 × 3 + 1 × 4 + 6 × 1 + 12 × 2

= 10 + 24 + 4 + 6 + 24

= Rs. 68

which is less than initial B.F.S.

Table 3

	2		7		4
5					
	3		3		7
		8			
	5		4		1
		1		6	
			6		2
2				12	

Step 8. Proceeding as in step 2, 3 and 4 we get the following table.

Table 4

		A_1			A_2			A_3		Available	ui
	2		2	6		7	3		4		
B_1		5								5.	1
	0			1			1				
	–1		3	3		3	0		7		
B_2					8					8	–2
	4			0			7				
	0		5	4		4	1		1		
B_3					1			6		7	–1
	5										
	1		1	5		6	2		2		
B_4		2			⊗			12		14	0
Required		7			9			18			
V_J		1			5			2			

Since all $d_{iJ} > 0$. Hence the B.F.S is optimal which is also unique.

Hence total transportation cost

$= 5 \times 2 + 8 \times 3 + 4 \times 1 + 6 \times 1 + 12 \times 2$

$= 10 + 24 + 4 + 6 + 24$

= Rs. 68.

Thus the solution of the given transportation problem is.

From source

B_1	to	A_1	Units	5
B_2	to	A_2	Units	8
B_3	to	A_2	Units	1
B_3	to	A_3	Units	6
B_4	to	A_1	Units	2
B_2	Units	A_3	to	4 Ans.

Example 3:

Solve the following transportation problem

Plant	*A*	*B*	*C*	*D*	*Available*
X	*10*	*22*	*10*	*20*	*8*
Y	*15*	*20*	*12*	*8*	*13*
Z	*20*	*12*	*10*	*15*	*11*
Required	*5*	*11*	*8*	*8*	*32/32*

Solut on:

Step 1. By Vogel's an initial B.F.S. of the given problem is given by following table 1.

Table 1

	A	B	C	D	Available
X	10 5	22	10 3	20	8
Y	15	20	12 5	8 8	13
Z	20	12 11	10 $\in$	15	11
Required	5	11	8	8	

Since the total number of allocation is 5 which is one less than m + n − 1 = 6.

Hence this solution is a degenerate solution. Now to resolve this degeneracy we allocate a very small amount ∈ to the cell (3, 3) getting 6 allocation at independent position.

Total transportation cost

$= 5 \times 10 + 3 \times 10 + 5 \times 12 + 8 \times 8 + 11 \times 12$

$= 50 + 30 + 60 + 64 + 132 =$ Rs. 336.

Step 2. In step 2 we will find u_i and v_j. For this we choose $u_2 = 0$.

$c_{23} = u_2 + v_3$; $\quad c_{24} = u_2 + v_4$

$12 = 0 + v_3$; $\quad 8 = 0 + v_4$

$\Rightarrow v_3 = 12$; $\quad \Rightarrow v_4 = 8$

Also $\quad c_{13} = u_1 + v_3 \quad c_{11} = u_1 + v_1 \quad c_{33} = u_3 + v_3$

$10 = u_1 + 12 \quad 10 = -2 + v_1 \quad 10 = u_3 + 12$

$\Rightarrow \quad u_1 = -2 \quad v_1 = 12 \quad \Rightarrow u_3 = -2$

$c_{32} = u_3 + v_2 \quad 12 = -2 + v_2 \quad \Rightarrow v_2 = 14.$

Step 3. Now we find the cell evaluations $u_i + v_j$ for each unoccupied cell (i, j) and enter at the upper left corner of the corresponding unoccupied cell.

Step 4. Then we find the cell evaluation $d_{ij} = c_{ij} - (u_i + v_j)$ (*i.e.*, the difference of the upper right) for each unoccupied cell (i, j) and enter at the lowest left corner of the corresponding unoccupied cell.

Thus we get the following table.

Table 2

	A		B		C		D		Available	ui
	10		10	12	22	−2		10	−6	20
X		5			3				18	−2
	0		10		12			26		
	12		15		14	20	0	12	−4	8
Y		5		8		13		0		
	3		6			12			12	
	10		20	12	12	−2		10	−6	15
Z		11					∈		11	−2
	10			0	12		21			
Required		5			11		8		8	
v_j	12			14	0		−4			

Since all $d_{il} \geq 0$, so the solution is optimal.

Thus the solution of the given problem is

X	to	A	Units	5
X	to	C	Units	3
Y	to	C	Units	5
y	to	D	Units	8
Z	to	B	Units	11

Example 4:

Determine an initial B.F.S to the following transportations problem by using the N.W. corner rule.

Destination

	A_1	B_1	C_1	D_1	E_1	
A	*2*	*11*	*10*	*3*	*7*	*4*
B	*1*	*4*	*7*	*2*	*1*	*8*
C	*3*	*9*	*4*	*8*	*12*	*9*
	3	*3*	*4*	*5*	*6*	

Solut on:

By 'N–W Corner rule' method we get the following B.F.S. of the given problem.

Table 1

	A_1	B_1	C_1	D_1	E_1	Supply
A	2	11	10	3	7	4
	3	1				
B	1	4	7	2	1	8
		2	4	2		
C	3	9	4	8	12	9
				3	6	
	3	3	4	5	6	

Total transportation cost $= 3 \times 2 + 1 \times 11 + 2 \times 4 + 4 \times 7 \times 2 \times 2 + 3 \times 8 + 6 \times 12$

$$= 6 + 11 + 8 + 28 + 4 + 24 + 72$$

$$= \text{Rs } 153$$

Example 5:

Solve the following transportation problem.

	A	*B*	*C*	*Available*
X	*7*	*3*	*4*	*2*
Y	*2*	*1*	*3*	*3*
Z	*3*	*4*	*6*	*5*
Demand	*4*	*1*	*5*	

Solut on:

By vogel' method an initial B.F.S of the given problem is given by following table.

Table 1

	A		B		C		
		7		3		4	
X					2		2
		2		1		3	
Y			1		2		3
		3		4		6	
Z	4				1		5
		4		1		5	

Total transportation cost

$= 2 \times 4 + 1 \times 1 + 2 \times 3 + 4 \times 3 + 1 \times 6 = 8 + 1 + 6 + 12 + 6 =$ Rs. 33.

Now finding the set u_i (1, 2, 3), v_j (J = 1, 2, 3) such that $c_{rs} = u_r + v_s$ for occupied cell and entering $u_i + v_j$ and d_{ij} in unoccupied cells, the table 2 giving the optimal solution as follows.

		A			B			C		Available	U_i
	!		7	2		3	4		4		
X								2		2	–2
	6			1			0				
	0		2	1		1	3		3		
Y					1			2		3	–3
	2			0			0				
	3		3	4		4	6		6		
Z		4						1		5	0
	0			0			0				
Demand	4			1			5				
v_J	3			4			6				

Since all $d_{iJ} > 0$ therefore above solution is optimal

X	to	C	Units	2
Y	to	B	Units	1
Y	to	C	Units	2
Z	to	A	Units	4
Z	to	C	Units	1

Transportation cost = Rs. 33.

Example 6:

Obtain an initial B.F.S to the following.

	S_1	S_2	S_3	S_4	*Availability*
A	*5*	*1*	*3*	*3*	*34*
B	*3*	*3*	*5*	*4*	*15*
C	*6*	*4*	*4*	*3*	*12*
D	*4*	*1*	*4*	*2*	*19*
	21	*25*	*17*	*17*	

Solution:

By vogel's method an initial B.F.S of the problem is given by following table.

Table 1

	S_1	S_2	S_3	S_4	
	5	1	3	3	
A		(25)	(9)		34
	3	3	5	4	
B	(15)				15
	6	4	4	3	
C	(4)		(8)		12
	4	1	4	2	
D	(2)			(17)	19
	21	25	17	17	

A	to	S_2	Units	25
A	to	S_3	Units	9
B	to	S_1	Units	15
C	to	S_1	Units	4
C	to	S_3	Units	8
D	to	S_1	Units	2
D	to	S_4	Units	7

Total Transportation cost

$= 25 \times 1 + 9 \times 3 + 15 \times 3 + 4 \times 6 + 8 \times 4 + 2 \times 4 + 17 \times 2$

$= 25 + 27 + 45 + 24 + 32 + 8 + 34 =$ Rs. 195.

Example 7:

Solve the following transportation problems

	D_1	D_2	D_3	*Supply*
O_1	2	4	1	40
O_2	6	3	2	50
O_3	4	5	6	20
O_4	3	2	1	30
O_5	5	2	5	10
Demand	50	60	40	

Solut on:

By penalty method we get following B.F.S of the given problem.

Table 1

	D_1		D_2		D_3		Supply
		2		4		1	
O_1	40						40
		6		3		2	
O_2			10		40		50
		4		5		6	
O_3	10		10				20
		3		2		1	
O_4			30				30
		5		2		5	
O_5			10				10
Demand	50		60		40		

Total transportation cost

$= 40 \times 2 + 10 \times 3 + 40 \times 2 + 10 \times 4 + 10 \times 5 + 30 \times 2 + 10 \times 2$

$= 80 + 30 + 80 + 40 + 50 + 60 + 20 =$ Rs. 360.

Now finding the set $U_i = (1, 2, 3)$, v_j ($j = 1, 2, 3, 4, 5$ such that $c_{rs} = u_r + v_s$ for occupied cells and entering $u_i + v_j$ and d_{ij} in unoccupied cells, the table 2 giving the optimal solution is as follows

Table 2

	2		2	3		4	2		1	u_i
		40						0		3
	0			1			–1			
	2		6	3		3	2		2	
					10			40		3
	4			0		0				
	4		4	5		5	4		6	
		10			10					5
	0			0		2				
	1		3	2		2	1		1	
					30					2
	2			0			0			
	1		5	2		2	1		5	
					10					2
	4			0			4			
v_J		–1			0			–1		

Since $d_{13} < 0$ so solution is not optimal.

Since mini $d_{iJ} = -$ (neg.), so we give maximum allocation to this cell from an occupied cell and make necessary changes in other allocation.

The new B.F.S is shown in table 3. For this B.F.S total transporting cost

$= 30 \times 2 + 10 \times 1 + 20 \times 3 + 30 \times 2 + 20 \times 4$

$+ 30 \times 2 + 10 \times 2$

$= 60 + 10 + 60 + 60 + 80 + 60 + 20 =$ Rs. 350.

Which is less than initial B.F.S.

Table 3

	2				1
30				10	
			3		2
		20		30	
	4				
20					
			2		
		30			
			2		
		10			

Proceeding as usual the fifth table as follows:

Table 4

		D_1			D_2			D_3		Supply	u_i
	2		2	2		4	1		1		
O_1		30						10		40	2
	0		2				0				
	3		6	3		3	2		2		
O_2					20			30		50	3
	3			0			0				
	4		4	4		5	–3		6		
O_3		20								20	4
	0			1			9				
	2		3	2		2	1		1		
O_4					30					30	2
	1			0			0				
	2		5	2		2	1		5		
O_5					10					10	2
	3			0			4				
Demand	50			60			40				
v_J		0			0			–1			

Since all d_j are positive therefore solution is optimal.

Now Transportation cost

$$= 30 \times 2 + 10 \times 1 + 20 \times 3 + 30 \times 2$$
$$+ 20 \times 4 + 30 \times 2 + 10 \times 2$$

$= 60 + 10 + 60 + 60 + 80 + 60 + 20$

$=$ Rs. 350.

Example 8:

Determine the optimum basic feasible solution to the following transportation problem:

	D_1	D_2	D_3	D_4	a_i
O_1	5	3	6	2	19
O_2	4	7	9	1	37
O_3	3	4	7	5	34
$b_j \rightarrow$	16	18	31	25	

Solut on:

Step 1. Using Vogel's method, the initial basic feasible solution is obtained as follows:

	D_1	D_2	D_3	D_4	
O_1	(5)	(3) 18	(6) 1	(2)	19
O_2	(4) 12	(7)	(9)	(1) 25	37
O_3	(3) 4	(4)	(7) 30	(5)	34
	16	18	31	25	

Total transportation cost $= 18.3 \quad 1.6 + 12.4 + 25.1 + 4.3 - 30.7$

$=$ Rs. 355.

Step 2. Now u_i (i = 1, 2 3) and v_j (j = 1, 2, 3, 4) are to be determined by means of the unit cost in the respective occupied cells only. For each occupied cell (r, s). $c_{rs} = r_r - v_s$.

Since all rows contain the same (maximum) number of allocations, so take any of the u_i (say u_3) equal to zero.

When $u_3 = 0$, $v_3 = 7$ (since $c_{33} = u_3 + v_3$; $c_{33} = 7$).

Similarly $\quad c_{31} = u_3 + v_1$

or $\quad 3 = 0 + v_1$ or $v_1 = 3$.

Again $\quad c_{21} = u_2 + v_1$

or $\quad 4 = u_2 + 3$

or $u_2 = 1.$

In the same way, $c_{24} = 1 = u_2 + v_4$ gives $v_4 = 0$;

$c_{13} = 6 = u_1 + v_3$ gives $u_1 = -1$

and $c_{12} = 3 = u_1 + v_2$ gives $v_2 = 4$.

and $\sum_{i=1}^{m} x_{ij} = b_j$

$$\text{or} \quad 0 = b_j - \sum_{i=1}^{m} x_{ij}, j = 1, 2, \ldots, n \qquad \ldots(3)$$

Multiplying (2) by u_i (i = 1, 2, .. ,m). (3) by v_j (j = 1, 2, ... ,n) and adding to (1). we have

$$Z = \sum_{i=1}^{m}\sum_{j=1}^{n} c_{ij}x_{ij} + \sum_{i=1}^{m} u_i\left(a_1 - \sum_{j=1}^{n} x_{ij}\right) + \sum_{j=1}^{n} v_j\left(b_j - \sum_{j=1}^{m} x_{ij}\right)$$

$$\text{or} \quad Z = \sum_{i=1}^{m}\sum_{j=1}^{n}[c_{ij} - (u_i + v_j)]x_{ij} + \sum_{i=1}^{m} u_i a_i + \sum_{j=1}^{n} v_j b_j \qquad \ldots(4)$$

But it is given that for each occupied cell (r, s) (cell with positive allocation)

$$c_{rs} = u_r + v_{s.} \qquad \ldots.(5)$$

Hence in the objective function (4), all the terms of positive allocations vanish as their coefficients are zero. Thus, for this feasible solution the value of the objective function (4) reduces to

$$Z = \sum_{i=1}^{m} u_i a_i + \sum_{j=1}^{n} v_j b_j \qquad \ldots(6)$$

Let us determine the cell evaluation for the empty cell (h, k). When we allocate one unit to this empty cell, the positive allocations become (m + n) in number and hence they become dependent in position. So a closed loop can be formed. Let the closed loop thus formed be as shown in the following figure.

Here cell (h, s), (r, s) and (r, k) are all occupied cells and hence

$c_{hs} = u_h + v_s$, $c_{rs} = u_r + v_s$, $c_{rk} = u_r + v_k$.

We have to decrease the individual allocations at (h, s) and (r, k) cells and increase at cell (r, s) by 1 unit to maintain the row and column sums. So the values of the individual allocations in these occupied cells are changed but in the objective function (4) they contribute nothing as their coefficients are necessarily zero. Thus, corresponding to this new solution [allocating 1 unit at cell (h, k)], the value of the objective function is given by

$$Z' = c_{hk} - (u_h + v_k) = \sum_{i=1}^{m} u_i a_i + \sum_{j=1}^{n} v_j b_j \qquad \text{...(7)}$$

Hence the cell evaluation is given by $d_{hk} = Z' - Z = c_{hk} - (u_h + v_k)$.

Thus, in general to each empty cell (i, j) the cell evaluation is given by

$$d_{ij} = c_{ij} - (u_i + v_j).$$

This completes the proof of the theorem.

Note: The above result is proved by considering a square (or rectangle) shaped loop. It can be generalized by considering a loop of an arbitrary shape connecting empty cell to the occupied cells.

Step 3. Now we find the evaluation $u_i + v_j$ for each unoccupied cell (i, j) and enter it at the upper right corner of the corresponding unoccupied cell.

Step 4. Then we find the cell evaluation $d_{ij} = c_{ij} - (u_i + v_j)$ for each unoccupied cell (i, j) and enter at the lower right corner of the corresponding unoccupied cell.

Thus we get the following table:

					$u_i \downarrow$
	(5) 2 3	(3) 18	(6) 1	(2) −1 3	−1
	(4) 12	(7) 5 2	(9) 8 1	(1) 25	1
	(3) 4	(4) 4 0	(7) 30	(5) 0 5	0
$v_j \rightarrow$	3	4	7	0	

Step 5. Since all $d_{ij} \geq 0$. The solution under test is optimal. But an alternative optimal solution will also exist as $d_{32} = 0$.

Thus the solution of the given problem is $x_{12} = 18$, $x_{13} = 1$, $x_{21} = 12$, $x_{24} = 25$, $x_{31} = 4$, $x_{33} = 30$ and minimum transportation cost = Rs. 355.

Example 9:

Find the optimum solution to the following transportation problem:

	D_1	D_2	D_3	D_4	Supply
O1	23	27	16	18	30
O2	12	17	20	51	40
O3	22	28	12	32	53
Demand	22	35	25	41	

Solut on:

Step 1. By Vogel's method an initial B.F.S. of the given problem is given in the following table:

	D_1	D_2	D_3	D_4	$a_i\downarrow$
O_1	(23)	(27)	(16)	(18) 30	30
O_2	(12) 5	(17) 35	(20)	(51)	40
O_3	(22) 17	(28)	(12) 25	(32) 11	53
b_j	22	35	25	41	

Total transportation cost = 18.30 + 12.5 + 17.35 + 22.17 + 12.25 + 32.11 = Rs. 2221.

Step 2. Now we determine a set of u_i and v_j such that for each occupied cell (r, s) $c_{rs} = u_r + v_s$.

For this we choose $u_3 = 0$ (since row 3 contains maximum number of allocations).

Since $c_{31} = 22 = u_3 + v_1$, $c_{33} = 12 = u_3 + v_3$, $c_{34} = 32 = u_3 + v_4$

$\therefore \quad v_1 = 22, v_3 = 12, v_4 = 32.$

Also $c_{21} = 12 = u_2 + v_1$,

$c_{22} = 17 = u_2 + v_2$, $c_{14} = 18 = u_1 + v_4$

$\therefore \quad u_2 = -10, v_2 = 27, u_1 = -14.$

Step 3. Now we find the evaluation $u_i + v_j$ for each unoccupied cell (i, j) and enter at the upper right corner of the corresponding unoccupied cell.

Step 4. Then we find the cell evaluation $d_{ij} = c_{ij} - (u_i + v_j)$ for each unoccupied cell (i, j) and enter at the lower right corner of the corresponding unoccupied cell.

Thus we get the following table:

					$u_i \downarrow$
	(23) 8 15	(27) 13 14	(16) –2 18	(18) 30	–14
	(12) 5	(17) 35	(20) 2 18	(51) 22 29	–10
	(22) 17	(28) 27 1	(12) 25	(32) 11	0
$v_j \rightarrow$	22	27	12	32	

Step 5. Since all d_{ij} for empty cell are > 0, the solution under test is optimal.

Thus the solution of the given problem is

$x_{14} = 30$, $x_{21} = 5$, $x_{22} = 35$, $x_{31} = 17$,

$x_{33} = 25$, $x_{34} = 11$ and minimum transportation cost = Rs. 2221.

Example 10:

Solve the following transportation problem in which cell entries represent unit cost:

	To			*Available*
	2	*7*	*4*	*5*
From	*3*	*3*	*1*	*8*
	5	*4*	*7*	*7*
	1	*6*	*2*	*14*
Required	*7*	*9*	*18*	

Solut on:

Step 1. Using Vogel's method, the initial B.F.S. is obtained as given below:

				a_i
	(2) 5	(7)	(4)	5
	(3)	(3)	(1) 8	8
	(5)	(4) 7	(7)	7
	(1) 2	(6) 2	(2) 10	14
$b_j \rightarrow$	7	9	18	

Total transportation cost = 2.5 + 1.8 + 4.7 + 1.2 + 6.2 + 2.10 = Rs..80.

Step 2. To test the solution for optimality, we find a set of u_i (i = 1, 2, 3, 4) and v_j (j = 1, 2, 3) such that for each occupied cell (r, s), $c_{rs} = u_{rs} = u_r + v_s$.

For this let us choose $u_4 = 0$ (as row 4 contains maximum number of allocations).

Now $c_{41} = 1 = u_4 + v_1$, $c_{42} = 6\ u_4 + v_2$, $c_{43} = 2 = u_4 + v_3$

$\therefore \quad v_1 = 1,\ v_2 = 6,\ v_3 = 2.$

Also $c_{11} = 2 = u_1 + v_1$, $c_{23} = 1 = u_2 + v_3$, $c_{32} = 4 = u_3 + v_2$

$\therefore \quad u_1 = 1,\ u_2 = -1,\ u_3 = -2.$

Step 3. Now we find the evaluation $u_i + v_j$ for each unoccupied cell (i, j) and enter it at the upper right corner of the corresponding unoccupied cell.

Step 4. Form the above table, we observe that the cell evaluation $d_{22} = -2$, is negative. Therefore, the solution under test is not optimal. The solution can be improved as shown in the next step.

Step 5. Since minimum d_{ij} is $d_{22} = -2$ (negative), so we allocate (say, θ) to the cell (2, 2) as much as possible.

Now when we have decided to include one more cell to the solution the occupied cells become dependent. Therefore, we first identify the loop joining cell (2, 2) with the occupied cells.

It is easily seen by the following rule that at the most θ = 2 units can be allocated from cell (4, 2) to cell (2, 2) still satisfying the row and column total and non-negativity restrictions on the allocations.

Step 6. Then we find the cell evaluation $d_{ij} = c_{ij} - (u_i + v_j)$ for each unoccupied cell (i, j) and enter it at the lower right corner of the corresponding unoccupied cell.

Thus, we get the following table:

				$u_i \downarrow$
	(2) 5	(7) 7 0	(4) 3 1	1
	(3) 0 3	(3) 5 −2	(1) 8	−1
	(5) −1 6	(4) 7	(7) 0 7	−2
	(1) 2	(6) 2	(2) 10	0
$v_j \rightarrow$	1	6	2	

Example 11:

There are three parties who supply and three who require the following quantities of coal:

Party 1	*14 tons*	*consumer A*	*6 tons*
Party 2	*12 tons*	*consumer B*	*10 tons*
Party 3	*5 tons*	*consumer C*	*15 tons*

The cost matrix is as follows:

	A	*B*	*C*
1	*6*	*8*	*4*
2	*4*	*9*	*3*
3	*1*	*2*	*6*

Find the schedule of transportation policy which minimizes the cost.

Solut on:

By matrix minima method the initial B.F.S. of the problem is as follows:

Destinations

		A	B	C	Available
		(6)	(8)	(4)	
Origins	1	1	10	3	14
		(4)	(9)	(3)	
	2			12	12
		(1)	(2)	(6)	
	3	5			5
Required		6	10	15	

Total transportations cost = 6.1 + 8.10 + 4.3 + 3.12 + 1.5 = Rs. 139.

Example 12:

Given the following data:

Destinations

		1	*2*	*3*	*Capacity*
	1	*2*	*2*	*3*	*10*
Sources	*2*	*4*	*1*	*2*	*15*
	3	*1*	*3*	×	*40*
Demand		*20*	*15*	*30*	

The cost of shipment form third source to the third destination is not known. How many units should be transported from sources to the destinations so that the total cost of transporting all the units to their destinations is a minimum.

Solut on:

Since the cost c_{33} is unknown, we assign a large cost, say M, to this cell. Then using Vogel's method an initial B.F.S. is obtained as shown in the table.

	1	2	3	$u_i \downarrow$
	(2)	(2)	(3)	
1			10	10
	(4)	(1)	(2)	
2			15	15
	(1)	(3)	M	
3	20	15	5	40
$b_j \rightarrow$	20	15	30	

To test the solution for optimality we have the following table:

								$u_i\downarrow$	
(2)		4–M	(2)		6–M	(3)			
							10	3–M	
		M–2			M–4				
(4)		3–M	(1)		5–M	(2)			
							15	2–M	
		M+1			M–4				
(1)			(3)			(M)			
	20			15			5		0
$v_j \rightarrow$	1			3			M		

Since M is very large, we observe that all the cell evaluations d_{ij} are ≥ 0. Hence, the current solution is optimum.

Thus the solution of the given problem is

$$x_{13} = 10,\ x_{23} = 15,\ x_{31} = 20,\ x_{32} = 15,\ x_{33} = 5.$$

Example 13:

Solve the following transportation problem:

	D_1	D_2	D_3	$a_i\downarrow$
O_1	7	4	0	5
O_2	6	8	0	15
O_3	3	9	0	9
$b_j \rightarrow$	15	6	8	

Solut on:

By 'North-West Corner Rule' the non-degenerate initial B.F.S. is obtained in the following table:

	D_1		D_2		D_3		
	(7)		(4)		(0)		
O_1		5					5
	(6)		(8)		(0)		
O_2		10		5			15
	(3)		(9)		(0)		
O_3				1		8	9
	15			6		8	

Now to test this solution for optimality we get the following table in usual manner:

	(7)		(4)		9	(0)		0	$u_i \downarrow$
		5							0
					–5			0	
	(6)		(8)			(0)		–1	
			10		5				–1
								1	
	(3)		7	(9)		(0)			
					1		8		0
			–4						
$v_j \rightarrow$		7			9			0	

Since all the cell evaluations are not ≥ 0, so the solution under test is not optimal. The largest negative cell evaluation is d_{12} = – 5, so we allocate as much as possible to the cell (1, 2).

5–θ	θ		5
10+θ	5–θ		15
	1	8	9
15	6	8	

(7)		(4)		(0)		
			5			5
(6)		(8)		(0)		
	15					15
(3)		(9)		(0)		
			1		8	9
	15		6		8	

Here maximum value of θ is obtained by usual rule:

$$\min [5 - \theta, 5, - \theta] = 0 \textit{ i.e.}, 5 - \theta = 0 \textit{ i.e.}, \theta = 5.$$

Since simultaneously two cells vacate, so the number of allocations becomes less than m + n – 1 *i.e.*, 5. Hence this is a degenerate solution we cannot apply the optimality test. A negligible quantity Δ may also be introduced in the independent cell (3, 1), although least cost independent cell is (2, 3).

(7)		(4)		(0)		
	5					5
(6)		(8)		(0)		
	15					15
(3)		(9)		(0)		
	Δ		1		89	9 + Δ = 9
15 + Δ = Δ			6	8		

To test this solution for optimality we have the following table:

(7)	–2	(4)			(0)		–5	$u_i \downarrow$
			5					–5
	9						5	
(6)		(8)		12	(0)		3	
	15							3
			–4			–3		
(3)		(9)		(0)				
	Δ		1			8		0
$v_j \rightarrow$	3		9		0			

Since all cell evaluations are not ≥ 0, so the solution under test is not optimal. The largest negative cell evaluation is $d_{22} = -4$, allocate as much as possible to the cell (2, 2).

	5	
15–θ	θ	
Δ+θ	1–θ	8

(7)		(4)		(0)	
			5		5
(6)		(8)		(0)	
	14		1		15
(3)		(9)		(0)	
	1			8	9
	15		6	8	

Here maximum value of θ is obtained by usual rule:

$$\min(15 - \theta,\ 1 - \theta) = 0 \text{ i.e., } 1 - \theta = 0 \text{ i.e., } \theta = 1.$$

To test this new improved solution for optimality we have the following table:

(7)		2	(4)			(0)		–1	$u_i \downarrow$
					5				–4
			5					1	
(6)			(8)			(0)		3	
	14			1					0
								–3	
(3)			(9)		5	(0)			
							8		
	1			4					–3
$v_j \rightarrow$	6			8			3		

Since all the cell evaluations are not ≥ 0, so the solution under test is not optimal. The largest negative cell evaluation is $d_{23} = -3$, allocate as much as possible to the cell (2, 3).

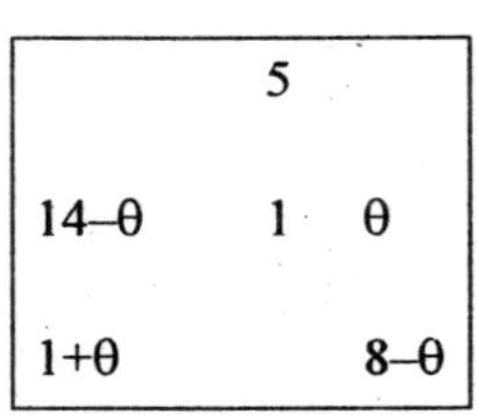

(7)	(4) 5	(0)	5
(6) 6	(8) 1	(0) 8	15
(3) 9	(9)	(0)	9
15	6	8	

Here maximum value of θ is obtained by usual rule:

$\min (14 - \theta, 8 - \theta) = 0$ *i.e.*, $8 - \theta$ or $\theta = 8$.

To test this new improved solution for optimality we have the following table:

				$u_i \downarrow$
	(7) 2 5	(4) 5	(0) –4 4	–4
	(6) 6	(8) 1	(0) 8	0
	(3) 9	(9) 5 8 4	(0) –3 3	–3
$v_j \rightarrow$	6	8	0	

Since all the cell evaluations for empty cells are positive, the solution under test is optimal. Thus the solution of the given problem is

$x_{12} = 5,\ x_{21} = 6,\ x_{22} = 1,\ x_{23} = 8,\ x_{31} = 9$

and minimum transportation cost = 4.5 + 6 + 6 + 8.1 + 0.8 +3.9 = Rs. 91.

Example 14:

A manufacturer wants to ship 8 loads of his product as shown in the table. The matrix gives the mileage from origin O to destination D. Shipping costs are Rs. 10 per load per mile. What shipping schedule should be used?

	D_1	*D_2*	*D_3*	*Available*
O_1	*50*	*30*	*220*	*1*
O_2	*90*	*45*	*170*	*3*
O_3	*250*	*200*	*50*	*4*
Required	*4*	*2*	*2*	

Solut on:

The initial B.F.S. by Vogel's method is obtained as follows:

	D_1	D_2	D_3	$a_i \downarrow$
O_1	(50) 1	(30)	(220)	1
O_2	(90) 3	(45)	(170)	3
O_3	(250)	(200) 2	(50) 2	4
$b_j \rightarrow$	4	2	2	

Since the total number of allocations is 4 which is one less than m + n – 1 = 5. Hence this solution is a degenerate solution. So the attempt to assign u_i and v_j values to the above table will not succeed. To resolve this degeneracy we allocate a very small amount Δ to some suitable cell. We allocate Δ to the cell (1, 2) getting 5 allocations at independent positions.

(50) 1	(30) Δ	(220)	1+Δ = 1
(90) 3	(45)	(170)	3
(250) 2	(200) 2	(50) 4	
4	2+Δ = 2	2	

To Test the Solution for Optimality

Now to test the solution for optimality we have the following table:

				$u_i \downarrow$
	(50) 1	(30) Δ	(220) –120 340	–170
	(90) 3	(45) 70 –25	(170) –80 250	–130
	(250) 220 30	(200) 2	(50) 2	0
$v_j \rightarrow$	220	200	50	

Since $d_{22} = -25 < 0$, so the solution under test is not optimal. Now we shall allocate to this cell (2, 2) as much as possible. Thus we take Δ from cell (1, 2) to cell (2, 2) and form the new table to check the solution for optimality.

	(50)		(30)		5	(220)		−145	$u_i\downarrow$
		1							−195
					25			365	
	(90)		(45)			(170)		−105	
		3		Δ				−155	
								275	
	(250)		245		(200)	(50)			
				2			2		0
			5						
$v_j \rightarrow$		245		200			50		

Since for empty cells all d_{ij} are > 0, so the solution under test is optimal.

Thus the solution of the given problem is

$$x_{11} = 1,\ x_{21} = 3,\ x_{32} = 2,\ x_{33} = 2$$

and minimum mileage = 50.1 + 90.3 + 200.2 + 50.2 = 820 *i.e.*, minimum cost = Rs. 8200.

Example 15:

Solve the following unbalanced transportation problem (symbols have their usual meanings).

	D_1	D_2	D_3	a_i
O_1	4	3	2	10
O_2	2	5	0	13
O_3	3	8	6	12
$b_j \rightarrow$	8	5	4	

Solut on:

Here $\Sigma a_i = 35$, $\Sigma b_j = 17$. Since Σa_i is greater than Σb_j, the problem is of unbalanced type. We convert this problem to a balanced one by introducing a fictitious destination D_4 with requirement 35 – 17 = 18 having all the transportation costs equal to zero. The balanced transportation table is given below:

	D_1	D_2	D_3	D_4	a_i
O_1	4	3	2	0	10
O_2	2	5	0	0	13
O_3	3	8	6	0	12
$b_j \rightarrow$	8	5	4	18	

Applying the Vogel's method in the usual manner, the initial B.F.S. is obtained as given below:

	D_1	D_2	D_3	D_4	$a_i \downarrow$
O_1	(4)	(3) 5	(2)	(0) 5	10
O_2	(2) 8	(5)	(0) 4	(0) 1	13
O_3	(3)	(8)	(6)	(0) 12	12
$b_j \rightarrow$	8	5	4	18	

This gives the transportation cost

$$= 3.5 + 0.5 + 2.8 + 0.4 + 0.1 + 0.12$$

$$= \text{Rs. } 31.$$

To test this solution for optimality we have the following table:

					$u_i \downarrow$
	(4) 2 2	(3) 5	(2) 0 2	(0) 5	0
	(2) 8	(5) 3 2	(0) 4	(0) 1	0
	(3) 2 1	(8) 3 5	(6) 0 6	(0) 12	0
$v_j \rightarrow$	2	3	0	0	

Since all $d_{ij} > 0$, so the solution under test is optimal.

Thus the solution of the given problem is

$$x_{12} = 5,\ x_{21} = 8,$$

$$x_{23} = 4$$

and min. cost = Rs. 31.

Example 16:

Solve the following transportation problem.

		To 1	2	3	Supply
From	1	2	7	4	5
	2	3	3	1	8
	3	5	4	7	7
	4	1	6	2	14
Demand		7	9	18	34/34

Solut on:

By Vogel's method we get following B.F.S of the given problem.

Table 1

2 (5)	7	4	5
3	3	1 (8)	8
5	4 (7)	7	7
1 (2)	6 (2)	2 (10)	14
7	9	18	

Total transportation cost = 5 × 2 + 8 × 1 + 2 × 1 + 2 × 6 + 10 × 21

= 10 + 8 + 2 + 12 + 20 + 28

= Rs. 80.

Now finding the set u_i (= 1, 2, 3, 4), v_j (= 1, 2, 3) such that $c_{rs} = u_r + v_s$ for occupied cells and entering $u_i + v_j$ and d_{ij} in unoccupied cells the table 2 giving the optimal solution is as follows :

Table 2

	2		2	7		7	3		4		u_i
		(5)									1
	0		0				1				
	0		3	5		3	1		1		
								(8)			–1
	3			–2			0				
	–1		5	4		4	0		7		
					(7)						–2
	6			0			7				
	1			1	6		6	2		2	
			(2)			(2)			10		0
	0			0			0				
v_j		1			6			2			

Since $d_{22} = -2 < 0$, so the solution is not optimal.

Since minimum d_{ij} is $d_{22} = -2$ (negative), so we give maximum allocation to this cell from an occupied cell and make the necessary changes in other allocations.

The new basic feasible is shown in table 3. For this B.S.F.

The total transportation cost

$$= 5 \times 2 + 2 \times 1 + 2 \times 3 + 7 \times 4 + 6 \times 1 + 12 \times 2$$

$$= \text{Rs. } 76.$$

which is less than that for the initial B.F.S.

Table 3

							Supply
		2		7		4	
	5						5
		3		3		1	
			2		6		8
		5		4		7	
			7				7
		1		6		2	
	2				12		14
Demand	7		9		18		

Proceeding as usual the fifth table as follows:

Table 4

										u_i
	2		2	5		7	3		4	
		5								1
	0			2			1			
	0		3	3		3	1		1	
					2			6		–1
	3			0			0			
	1		5	4		4	2		7	
					7					0
	4		0				5			
	1		1	4		6	2		2	
		2						12		0
	0			2			0			
v_J		1			4			2		

Since all $d_{ij} > 0$. Here B.F.S. is an optimal solution which is also unique.

Thus, the solution of given transportation problem is

From Source 1 transport 5 units to destination 1.

" " 2 " 2 and 6 Units " " 2&3.

" " 3 " 7 Units to " " 2

" " 4 " 2 and 12 Units to " 1&3

the total transportation cost = Rs. 76. Ans.

Example 17:

A company has four plants P_1, P_2, P_3, P_4 from which it supplies to three markets M_1, M_2, M_3. Determine the optimal transportation plan from the following data giving the plant to market shifting costs, quantities available at each plant and quantities required at each marked.

Market↓ \ Plant	P_1	P_2	P_3	P_4	Required at Market
M_1	19	14	23	11	11
M_2	15	16	12	21	13
M_3	30	25	16	39	19
Available at Plant	6	10	12	15	43

Solut on:

By Vogel's method an initial B.F.S. at the given problem is given by the following table.

Table 1

	19	14	23	11 11	Required 11
	15 6	16 3	12	21 4	13
	30	25 7	16 12	39 19	
Available	6 10	12	15		

The total transportation cost

$= 6 \times 15 + 3 \times 16 + 7 \times 25 + 12 \times 16 + 11 \times 11 + 4 \times 21$

$=$ Rs. 710.

Now finding the set $u_i = (1, 2, 3)$, v_J $(J = 1, 2, 3, 4)$ such that $c_{rs} = u_r + v_s$ for occupied cells and entering $u_i + v_J$ and d_{iJ} in unoccupied cell the table 2 giving the optimal solution is as follows.

Table 2

	5 19 14	6 14 8	–3 23 26 0	11 11 11	u_i –10
	15 15 6 0	16 16 3 0	7 12 5	21 21 4 0	0
	24 30 6	25 25 7 0	16 16 12	30 39 9 9	
v_J	15	16	7	21	

Since all $d_{ij} > 0$. Hence B.F.S. is an optimal solution and the optimal transportation cost

$=$ Rs. 710.

Thus, the solution of the given transportation problem is.

From source 1 transport 11 Units to destination 4.

" " 2 " 6, 3 and 4 Units to destination 1, 28 4 respectively

" " 3 7 and 12 Units to destinations 2 and 3 repetitively.

EXERCISE

1. Explain how it is tested whether a B.F.S. of a transportation problem is optimal or not.

2. How does the problem of degeneracy arise in a transportation problem?

3. Explain how to solve the degeneracy in transportation problems.

4. Explain briefly the step-wise description of the computational procedure for solving the transportation problem.

5. Obtain an initial basic feasible solution to the following transportation problem. Is this solution an optimal solution? If not, obtain the optimal solution.

	W_1	W_2	W_3	W_4	a_i
F_1	19	30	50	10	7
F_2	70	30	40	60	9
F_3	40	8	70	20	18
$b_j \rightarrow$ 5	8	7	14		

6. Solve the transportation problem:

	D_1	D_2	D_3	D_4	Available
O_1	1	2	1	4	30
O_2	3	3	2	1	50
O_3	4	2	5	9	20
Required	20	40	30	10	100 Total

7. Determine the optimum basic feasible solution to the following transportation problem:

	D_1	D_2	D_3	D_4	Capacity
O_1	1	2	3	4	6
O_2	4	3	2	0	8
O_3	0	2	2	1	10
Demand	4	6	8	6	

where O_i and D_j denote ith destination respectively.

8. Is $x_{13} = 50$, $x_{14} = 20$, $x_{21} = 55$, $x_{31} = 30$, $x_{32} = 35$, $x_{34} = 25$ an optimum solution of the following transportation problem?

					Available units
	6	1	9	3	70
	11	5	2	8	55
	10	12	4	7	90
Required units	85	35	50	45	

If not, modify it to obtain an optimum basic feasible solution.

9. A company has four plants P_1, P_2, P_3, P_4 from which it supplies to three markets M_1, M_2, M_3. Determine the optimal transportation plan from the following data giving the plant to market shifting costs, quantities available at each plant and quantities required at each market.

	P_1	P_2	P_3	P_4	Required
M_1	19	14	23	11	11
M_2	15	16	12	21	13
M_3	30	25	16	39	19
Available	6	10	12	15	

10. Solve the transportation problem where all entries are unit costs.

	D_1	D_2	D_3	D_4	D_5	a_i
O_1	73	40	9	79	20	8
O_2	62	93	96	8	13	7
O_3	96	65	80	50	65	9
O_4	57	58	29	12	87	3
O_5	56	23	87	18	12	5
$b_j \rightarrow$	6	8	10	4	4	

11. The following table gives the cost for transporting material from supply points A, B C and D to demand points E, F, G, H and J.

		To				
		E	F	G	H	J
	A	8	10	12	17	15
	B	15	13	18	11	9
From	C	14	20	6	10	13
	D	13	19	7	5	12

The present allocation is as follows:

A to E 90; A to F 10; B to F 10; C to G 50; C to J 120; D to H 210; D to J 70.

Check if this allocation is optimum. If not, find an optimum schedule.

12. Solve the following transporting problem for minimum cost:

		To				
		I	II	III	IV	
	A	15	10	17	18	2
From	B	16	13	12	13	6
	C	12	17	20	11	7
		3	3	4	5	

13. Solve the following problem:

			To	Supply	
	21	16	25	13	11
From	17	18	14	23	13
	32	27	18	41	19
Demand	6	10	12	15	43 Total

14. Given below the unit cost array with supplies a_i; i = 1, 2, 3 and demand b_j; j = 1, 2, 3, 4.

	8	10	7	6	50	$a_i \downarrow$
	12	9	4	7	40	
	9	11	10	8	30	
$b_j \rightarrow$	25	32	40	23		

Find the optimal solution to the above problem.

15. A company has three plants A, B and C and three warehouses X, Y and Z. Number of units available at the plants is 60, 70 and 80 respectively. Demands at X, Y and Z are 50, 80 and 80 respectively. Unit costs of transportation are as follows:

	X	Y	Z
A	8	7	3
B	3	8	9
C	11	3	5

What would by your transportation plan? Give minimum distribution cost.

16. The cost-requirement table for the transportation problem is given below:

		W_1	W_2	W_3	W_4	W_5	Available
From	F_1	4	3	1	2	6	40
	F_2	5	2	3	4	5	30
	F_3	3	5	6	3	2	20
	F_4	2	4	4	5	3	10
Required		30	30	15	20	5	

Obtain the optimal solution of the problem.

17. Solve the following transportation problem (cell entries represent unit cost):

							Available
	5	3	7	3	8	5	3
	5	6	12	5	7	11	4
	2	1	2	4	8	2	2
	9	6	10	5	10	9	8
Required	3	3	6	2	1	2	17

18. Obtain an optimum B.F.S. to the following degenerate transportation problem:

			To	Available
From	7	3	4	2
	2	1	3	3
	3	4	6	5
Demand	4	1	5	

19. Solve the following cost minimizing transportation problem:

(i)

	D_1	D_2	D_3	D_4	Available
O_1	5	3	6	5	15
O_2	10	7	12	4	11
O_3	7	5	8	4	13
Demand	8	12	13	6	

(ii)

	D_1	D_2	D_3	D_4	D_5	Available
O_1	5	5	6	4	2	9
O_2	6	9	7	8	5	13
O_3	5	6	4	6	3	9
Demand	3	7	8	5	8	

20. Consider the following unbalanced transportation problem:

To

		1	2	3	Supply
	1	5	1	7	10
From	2	6	4	6	80
	3	3	2	5	15
Demand	75	20	50		

Since there is not enough supply, some of the demands at these destinations may not by satisfied. Suppose there are penalty costs for every unsatisfied demand unit which are given by 5, 3 and 2 for destination 1, 2 and 3 respectively. Find the optimal solution.

21. Solve the following transportation problems.

Ware house Factory↓	W_1	W_2	W_3	W_4		Factory Capacity
	F_1	10	30	50	10	7
	F_2	70	30	40	60	9
	F_3	40	8	70	20	18
Ware house Requirement	5	8	7	14		

22.

	D_1	D_2	D_3	D_4	D_5	D_6	a_i
O_1	9	12	9	6	9	10	5
O_2	7	3	7	7	5	5	6
O_3	6	5	9	11	3	11	2
O_4	6	8	11	2	2	11	9
b_J	4	4	6	2	4	2	

23. A manufacturer wants to ship 8 loads of his product as shown below. The matrix gives the mileage from origin 0 to the destination

O ↓ D →	O	A	B	C	Available
X	50	30	220	1	
Y	90	45	170	3	
Z	250	200	50	4	
Required	4	2	2	8	

Shipping cost are Rs. 10 per load male. What shipping schedule should be used.

24. Solve the following T.C. problems.

(i)

	A	B	C	a_i
I	10	9	8	8
II	10	7	10	7
III	11	9	7	9
IV	12	14	10	4
b_J	10	10	8	

(ii)

	W_1	W_2	W_3	W_4	Capacity
F_1	95	105	80	15	12
F_2	115	180	40	30	7
F_3	155	180	95	70	5
Requirement	5	4	4	11	

25. Explain the difference between a transportation problem and an assignment problem.

26. Write a short note on transportation problem.

27. What is Transportation Problem? Give the mathematical formulation of transportation problem.

28. Explain the difference between a transporting problem and an assignment problem.

29. If all the sources are emptied and all the destinations are filled show that $\sum a_i = \sum b_j$ is a necessary and sufficient condition for the existence of a feasible solution to a transportation problem.

30. Prove that there are only m + n – 1 independent equations in a transportation problem, m and n being the no. of origins and destinations respectively and that any one equation can be dropped as the redundant equation.
31. Explain the north-west corner rule for obtaining an initial basic feasible solution of a transportation problem.
32. Explain the lowest cost entry method for obtaining an initial basic feasible solution of a transportation problem.
33. Explain Vogel's Approximation Method of solving a transportation problem.
34. Use north-west corner rule to determine an initial basic feasible solution to the following transportation problem:

(i) To

		I	II	III	IV	Supply
	A	13	11	15	20	2
Form	B	17	14	12	13	6
	C	18	19	15	12	7
Demand		3	3	4	5	

(ii) Destination

		D_1	D_2	D_3	D_4	Supply
	O_1	6	4	1	5	14
Origin	O_2	8	9	2	7	16
	O_3	4	3	6	2	5
Demand		6	10	15	4	

35. Determine an initial B.F.S. to the following transportation problem by using the north-west corner rule:

Destination

		I	II	III	IV	V	Supply
	A	2	11	10	3	7	4
	B	1	4	7	2	1	8
Origin	C	3	9	4	8	12	9
Demand		3	3	4	5	6	

36. Using 'lowest cost entry method' find the initial B.F.S. of the following transportation problem:

(i) Destinations

		A	B	C	D	Supply
	I	1	5	3	3	34
Origins	II	3	3	1	2	15
	III	0	2	2	3	12
	IV	2	7	2	4	19
Demand	21	25	17	17		

(ii) Destination

		D_1	D_2	D_3	D_4	Capacity
	O_1	1	2	3	4	6
Origin	O_2	4	3	2	0	8
	O_3	0	2	2	1	10
Demand		4	6	8	6	

37. Obtain an initial B.F.S. to the following transportation problem using Vogel approximation method:

To

		I	II	III	IV	Available
	A	5	1	3	3	34
	B	3	3	5	4	15
From	C	6	4	4	3	12
	D	4	1	4	2	19
Requirement		21	25	17	17	

38. Determine an initial B.F.S. to the following transportation table using (i) matrix minima method (ii) Vogel's approximation method:

Destination

		D_1	D_2	D_3	D_4	Supply
	O_1	1	2	1	4	30
Origin	O_2	3	3	2	1	50
	O_3	4	2	5	9	20
Demand		20	40	30	10	100

39. Find the initial basic feasible solution of the following transportation problem using (i) north-west corner rule (ii) matrix minima method (iii) Vogel's approximation method:

Warehouse

		W_1	W_2	W_3	W_4	Capacity
	F_1	19	30	50	10	7
Factory	F_2	70	30	40	60	9
	F_3	40	8	70	20	18
Requirement		5	8	7	14	

40. To

		W_1	W_2	W_3	W_4	W_5	Available
	F_1	3	4	6	8	9	20
	F_2	2	10	1	5	8	30
From	F_3	7	11	20	40	3	15
	F_4	2	1	9	14	16	13
Required		40	6	8	18	6	78

41. To

	A	B	C		Available
	X	7	3	4	2
From	Y	2	1	3	3
	Z	3	4	6	5
Demand		4	1	5	10

3

Mathematical Formulation of Assignment Problem

INTRODUCTION

The assignment problem of problem arises because available resources such as men, machines, etc. have varying degrees of efficiency for performing different activities. Therefore, cost, profit or time of performing the different activities is different. Thus, the problem is: *How should the assignments be made so as to optimize the given objective.* Some of the problems where the assignment technique may be useful are: Assignment of workers to machines, salesmen to different sales areas, clerks to various checkout counters, classes to rooms, etc. An assignment problem is a particular case of transportation problem where the objective is to assign a number of resources to an equal number of activities so as to minimize total cost or maximize total profit of allocation.

Mathematically an assignment problem can be stated as follows:

Minimize the total cost

$$Z = \sum_{i=1}^{n}\sum_{j=1}^{n} c_{ij} x_{ij}$$

where x_{ij} = {1, if ith person is assigned to the jth job 0, if not subject to the conditions

(i) $\sum_{j=1}^{n} x_{ij} = 1$, (only one job is done by the person, i = 1, 2, ... ,n)

(ii) $\sum_{i=1}^{n} x_{ij} = 1$, (only one person should be assigned to the jth job, j = 1, 2, ... ,n)

Remark: Since x_{ij} = 0 or 1 for all i and j exactly one variable in each row or column constant |s| (and other variables is zero) the structure of the problem is an nontransportation array and such three are 2n – 1 basic variables of these exactly n has the value 1 and the remaining n – 1, the value 0. Thus, the assignment problem is heavily degenerate and it will be frustrating to attempt to solve it by transportation method.

ROUTING PROBLEM

He knows the distances (or time or cost) of journey between every pair of cities allotted to him. His problem is to choose such a rout which starts from his home city, passes through each city once and only once and returns this home city in the shortest possible distance (or in least time or at least cost). Suppose a salesman wants to visit a certain number of cities.

The above problem may be classified in two forms:

(i) Symmetrical

If the distance (or time or cost) between every pair of cities is independent of the direction of journey the problem is said to be symmetrical.

(ii) Asymmetrical

If for one or more pairs of cities, the distance (or time or cost) changes with the direction, the problem is said to be asymmetrical. For example, it takes longer time while going up-hill from city A to B instead of coming down hill from city B to A. Similarly flying from east to west usually takes longer time than from West to East on account of prevailing winds.

Further, we note that for two cities there is only one possible route *i.e.*, there is no choice.

In case of three cities, say A, B and C, one of them (say A) is the home base, there are two possible routes: A → B → C and A → C → B. For four cities there are 3! = 6 possible routes. In general, there are (n – 1)! possible routes if there are n cities. Thus, practically it is impossible to find the best route by trying each one. That is why the travelling salesman problem is considered as a puzzle by the mathematicians. The best procedure to solve the problem is as if it were an assignment problem. We formulate the problem of travelling salesman in the form of an assignment problem with the additional restriction on his choice of route.

NATURE OF THE ASSIGNMENT PROBLEMS

Let there be n jobs to be performed and for doing these jobs n persons are available.

Assume that each person can do each job at a time, though with varying degree of efficiency.

Let c_{ij} be the cost (payment) of assigning ith person to the jth job. Then the problem is to find an assignment (which job should be assigned to which person) of that the total cost for performing all jobs is minimum.

The above assignment problem can be stated in the form of $n \times n$ matrix $[c_{ij}]$ of real numbers called cost *matrix* or effectiveness *matrix* as follows:

Cost Matrix

Jobs

		1	2	3	...	j	...	n
	1	c_{11}	c_{12}	c_{13}	...	c_{1j}	...	c_{1n}
	2	c_{21}	c_{22}	c_{23}	...	c_{2j}	...	c_{2n}
	3	c_{31}	c_{32}	c_{33}	...	c_{3j}	...	c_{3n}
	⋮	⋮	⋮	⋮		⋮		⋮
	⋮	⋮	⋮	⋮		⋮		⋮
	⋮	⋮	⋮	⋮		⋮		⋮
Persons	i	c_{i1}	c_{i2}	c_{i3}	...	c_{ij}	...	c_{in}
	⋮	⋮	⋮	⋮		⋮		⋮
	⋮	⋮	⋮	⋮		⋮		⋮
	⋮	⋮	⋮	⋮		⋮		⋮
	n	c_{n1}	c_{n2}	c_{n3}	...	c_{nj}	...	c_{nn}

ASSIGNMENT PROBLEM (MAXIMAL)

A problem of minimizing the total cost. Sometimes the assignment problem deals with the maximization of an objective function rather than to minimize it. For example, the problem may be to assign persons to jobs in such a way that the expected profit is maximum.

This problem may be solved easily by first converting it to a minimization problem and then we can use the usual procedure of assignment algorithm. This conversion can be very easily done by modifying the given profit matrix to the cost matrix in either of the two ways.

(i) *Select the greatest element of the given profit matrix and then subtract each element of the matrix from this greatest element to get the modified matrix.*

For, if $[c_{ij}]$ is the given profit matrix and c_{rk} is the greatest element of this matrix then the modified matrix will be $[c_{ij}']$, where $c_{ij}' = c_{rk} - c_{ij}$. It can be shown that if $x_{ij} = x_{ij}$ maximizes $Z = \Sigma\Sigma c_{ij}\, x_{ij}$, then $x_{ij} = X_{ij}$ minimizes the function $Z' = \Sigma\Sigma c_{ij}'\, x_{ij}$, where $c_{ij}' = c_{rk} - c_{ij}$. It follows from the relation

$$Z' = \Sigma\Sigma c_{ij}'\, x_{ij} = \Sigma\Sigma\,(c_{rk} - c_{ij})\, x_{ij} = \Sigma\Sigma c_{rk}\, x_{ij} - \Sigma\Sigma c_{ij}\, x_{ij} = nc_{rk} - Z.$$

(ii) *Place minus sign before each element of the profit matrix to get the modified matrix.*

In this case if $[c_{ij}]$ is the given profit matrix then the modified matrix will be $[c_{ij}']$ where $c_{ij}' = -\,c_{ij}$. It can be shown that if $x_{ij} = X_{ij}$ maximizes $Z = \Sigma\Sigma c_{ij}\, x_{ij}$ then $x_{ij} = X_{ij}$ minimizes $Z' = \Sigma\Sigma c_{ij}'\, x_{ij}$.

ASSIGNMENT PROBLEM (UNBALANCED)

In case the number of tasks (jobs) is not equal to the number of facilities (persons), the assignment problem is called an *unbalanced assignment problem.* Thus the cost matrix of an unbalanced assignment problem is not a square matrix. For the solution of such problem we add the dummy (fictitious) rows or columns to the given matrix with zero costs to form it a square matrix. Then the usual assignment algorithm can be applied to this resulting baseness assignment problem.

FUNDAMENTAL THEOREMS

Now, we shall prove two important theorems on which the solution to an assignment problem is fundamentally based.

Theorem 1:

(Reduction Theorem). *In an assignment problem if we add (or subtract) a constant to every element of a row (or column) of the cost matrix $[c_{ij}]$, then an assignment which minimizes the total cost for one matrix also minimizes the total cost for the other matrix.*

Or

Mathematical statement of reduction theorem.

If $x_{ij} = X_{ij}$ *minimizes* $Z = \sum_{i=1}^{n}\sum_{j=1}^{n} c_{ij}x_{ij}$ *over all* $x_{ij} = 0$ *or* 1 *such that*

$\sum_{i=1}^{n} x_{ij} = 1$, $\sum_{j=1}^{n} x_{ij} = 1$, *then* $x_{ij} = X_{ij}$ *also minimizes*

$Z' = \sum_{i=1}^{n}\sum_{j=1}^{n} c_{ij}'x_{ij}$ *where* $c_{ij}' = c_{ij} \pm a_i \pm b_j$; b_i ; a_i, b_j *are some real numbers for* $i, j = 1, 2, \ldots, n$.

Proof:

We have $Z' = \sum_{i=1}^{n}\sum_{j=1}^{n} c_{ij}'x_{ij}$

$$= \sum_{i=1}^{n}\sum_{j=1}^{n}(c_{ij} \pm a_i \pm b_j)x_{ij}$$

$$= \sum_{i=1}^{n}\sum_{j=1}^{n} c_{ij}x_{ij} \pm \sum_{i=1}^{n}\sum_{j=1}^{n} a_i x_{ij} \pm \sum_{i=1}^{n}\sum_{j=1}^{n} b_j x_{ij}$$

$$= Z \pm \sum_{i=1}^{n} a_i \sum_{j=1}^{n} x_{ij} \pm \sum_{j=1}^{n} b_j \sum_{i=1}^{n} x_{ij} \qquad \left[\because Z = \sum_{i=1}^{n}\sum_{j=1}^{n} c_{ij}x_{ij}\right]$$

$$= Z \pm \sum_{i=1}^{n} a_i \times 1 \pm \sum_{j=1}^{n} b_j.1 \qquad \left[\because \sum_{i=1}^{n} x_{ij} = 1 = \sum_{j=1}^{n} x_{ij}\right]$$

$$= Z \pm \sum_{i=1}^{n} a_i \pm \sum_{j=1}^{n} b_j.$$

Since terms $\sum_{i=1}^{n} a_i, \sum_{j=1}^{n} b_j$ are independent of x_{ij}'s, if follows that Z' is minimized whenever Z is minimized and conversely.

Hence, $x_{ij} = X_{ij}$ which minimizes Z well also minimize Z'.

Theorem 2:

If all $c_{ij} \geq 0$ and there exists a solution $x_{ij} = X_{ij}$ which satisfies

$$\sum_{i=1}^{n}\sum_{j=1}^{n} c_{ij}x_{ij} = 0,$$

then this solution is an optimal solution for the problem (i.e., minimizes the objective function).

Proof:

Since all $c_{ij} \geq 0$ and all $x_{ij} \geq 0$, the objective function $Z = \sum_{i=1}^{n}\sum_{j=1}^{n} c_{ij}x_{ij}$ cannot be negative. The minimum possible value that Z can attain is 0.

Hence the solution $x_{ij} = X_{ij}$ for which $\sum_{i=1}^{n}\sum_{j=1}^{n} c_{ij}x_{ij} = 0$ is an optimal solution.

ASSIGNMENT ALGORITHM (HUNGARIAN ASSIGNMENT METHOD)

Various steps of the computational procedure for obtaining an optimal assignment are as follows.

Step 1. Modify cost matrix by subtracting the minimum element of each row of the cost matrix, from all the elements of respective rows. Further, modify the resulting matrix by subtracting the minimum element of each column from all the elements of the respective columns. These operations create zeros.

Step 2. In the modified matrix search for an optimal assignment.

Starting with row 1 or the matrix obtained in step 1, examine rows successively until a row with exactly one zero element is found. Mark '( )' at this zero as an assignment will be made there. Mark '×' at all other zeros if lying in the column containing the assigned zero. This eliminates the possibility of marking further, assignments in that column. Continue in this manner until all the rows have been examined.

When the set of rows has been completely examined, an identical procedure is applied successively to columns. Starting with column 1, examine all the columns until a column containing exactly one unmarked zero is found. Then make an assignment in that position [indicated by '( )'] and mark '×' at all zeros in the row containing this marked zero. Proceed in this way until the last column is examined.

Continue the above operations on rows and columns successively until we reach to any of the two situations

(i) all the zeros have been marked '( )' or '×'

(ii) the remaining zeros lie at least two in each row and column.

In situation (i), we have a maximal assignment (assignment as much as we can) and in situation (ii) still we have some zeros to be treated. To work with such situations of zeros there is again an algorithm, complicated enough. But to avoid this highly complicated algorithm we use the trial and error method to break up such ties of zeros.

Now there are tow possibilities:

(i) If there is an assignment in every row and every column (*k*, total number of marked '( )' zeros is exactly n), then we have obtained a complete optimal assignment plan.

(ii) If every row and every column do not contain an assignment (*i.e.*, total number of marked '( )' zeros is less than n), then we shall modify the cost matrix by creating some more zeros in it.

Step 3. If in step 2 every row and every column of the matrix do not contain assignment then draw the minimum number of horizontal and vertical lines to cover all the zeros at least once in the resulting matrix.

Rule to Draw Minimum Number of Lines

(i) Mark (√) all rows that do not have any marked '( )' zero.

(ii) Mark (√) columns which have zeros in marked rows.

(iii) Mark (√) rows (not already marked) which have assignments in marked columns.

(iv) Repeat steps (ii) and (iii) until the chain of marking ends.

(v) Draw lines through unmarked rows and through marked columns.

This will give us the minimal system of lines.

Notes:

1. The lines thus drawn (horizontal and vertical both) are the minimum number of lines to pass through all the zeros of the matrix. It can be shown that the minimum number of lines required to pass through all the zeros of the matrix is the same as the maximum number of assigned independent zeros of the matrix.

   Thus, if the number of lines is exactly n, then the complete assignment plan is obtained wile if the number of lines is less than n, then the complete assignment is not possible.

2. These lines cover all the zeros and each line passes through one and only one marked zero (assignment). If there are two marked '( )' zeros in a row then it follow that we are assigning two jobs to one person which is a violation of the hypothesis. Thus, no line passes through more than one marked '( )' zero.

Step 4. Select the smallest of the elements that do not have a line though them, subtract if from all the elements that do not have a line through them, add it to every element that lies at the intersection two lines and leave the remaining elements of the matrix unchanged. In the modified matrix number of zeros are increased (never decrease) than that in step 2. Now apply the step to this new matrix. If still a complete optimal assignment is not possible in this matrix, then repeat steps 3 and 4 iteratively. Continue the process until minimum number of lines be n.

Thus, exactly one marked '( )' zero in each row and each column of the matrix is obtained. The assignment corresponding to these marked '( )' zero in each column of the matrix is obtained. The assignment corresponding to these marked '( )' zeros will give the optimal assignment.

Note: The procedure of subtracting the minimum element of all uncovered elements from all such elements and adding this minimum element to the elements placed at the intersection does not change the optimum solution *i.e.*, the two matrices will have the same optimal solution. For the above mentioned two operations of addition and subtraction are the resultant of operations of subtracting the above chosen minimum element from the uncrossed rows and adding it to all the crossed columns and such operations do not change the optimum solution.

SENSITIVITY ANALYSIS IN ASSIGNMENT PROBLEMS

There is very little scope for sensitivity analysis in assignment problem because of its structure. Modest alterations in the conditions (such as one being able to do two jobs) can be considered by repeating the man's row and adding a dummy a column to square up the matrix. The problem of not assigning a particular job (ith) to particular facility (jth) can be solved by taking a very large cost (∞) of this assignment.

Also, the addition of a constant throughout any row or column does not affect the optimal solution of the assignment problem.

FORMULATION OF A TRAVELLING SALESMAN PROBLEM AS ASSIGNMENT PROBLEM

Let x_{ij} = 1, if the salesman goes directly from city A_j, to c_{ij} to A_j and zero otherwise. Also, let c_{ij} be the distance (or time or cost) from city A_i to city A_j. Then our problem is to minimize $Z = \sum\sum c_{ij}\, x_{ij}$ with one additional restriction that the x_{ij}'s must be so chosen that no city is visited i_j twice until the tour of all the cities is completed.

In particular, he cannot go directly from city A_i to A_i itself. To avoid this possibility in the minimization process we adopt the convention $c_{ij} = \infty$ so that x_{ij} can never be unity when i = j. Also we note that only one x_{ij} = 1 for each value of i and j. The distance (or time or cost) matrix for this problem is given in the following Table:

To

	A_1	A_2		...	A_n
	A_1	∞	c_{12}	...	c_{1n}
From	A_2	c_{21}	∞	...	c_{2n}
	A_n	c_{n1}	c_{n2}	...	∞

We can omit the variable x_{ij} from the problem specification. Our problem is to determine a set of n elements of this matrix, one in each row and one in each column, so as to minimize the sum of the elements determined above.

Note: A problem similar to travelling salesman arises when n items say A_i, i = 1, 2, ... n, are to be produced on a machine in continuation, given that c_{ij} (i, j = 1, 2, ... n) is the setup cost of the machine when item A_i is followed by A_j. Here, two additional restrictions are imposed. One restriction is that we do not follow A_i again by A_i. The other restriction is that we do not produce an item again until all items are produced once.

SOLUTION PROCEDURE

The problem could be solved by assignment technique. In some cases (violation of additional restriction) we use the method of numeration by assignment the next minimum element of matrix in place of zero.

The following example will make the procedure clear.

DIFFERENT METHODS OF ASSIGNMENT PROBLEM

An assignment problem can be solved by the following four methods:

1. Enumeration method
2. Simplex method
3. Transportation method
4. Hungarian method.

Enumeration Method

In this method, a list of all possible assignments among the given resources (men, machines, etc.) and activities (jobs, sales areas, etc.) is prepared. Then an assignment involving the minimum cost (or maximum profit), time or distance is selected. If two or more assignments have the same minimum cost (or maximum profit), time or distance, the problem has multiple optimal solutions.

In general, if an assignment problem involves n workers/jobs, then there are in total n! possible assignments. For example, for an $n = 5$ workers/jobs problem, we have to evaluate a total of 5! or 120 assignments. However, when n is large, the method is unsuitable for manual calculations. Hence, this method is suitable only for small n.

Simplex Method

Since each assignment problem can be formulated as a 0 or 1 integer linear programming problem, such a problem can also be solved by the simplex method. As can be seen in the general mathematical formulation of the assignment problem, there are $n \times n$ decision variables and $n + n$ or In equalities. In particular, for a problem involving 5 workers/jobs, there will be 25 decision variables and 10 equalities. It is, again, difficult to solve manually.

Transportation Method

Since an assignment problem is a special case of the transportation problem, it can also be solved by transportation methods discussed. However, every basic feasible solution of a general assignment problem having a square payoff matrix of order n should have m + n – 1 = n + n – 1 = 2n – 1 assignments. But due to the special structure of this problem, any solution cannot have more than n assignments. Thus, the assignment problem is inherently degenerate. In order to remove degeneracy, (n – 1) number of dummy allocations (deltas or epsilons) will be required in order to proceed with the transportation method. Thus, the problem of degeneracy at each solution makes the transportation method computationally inefficient for solving an assignment problem.

Hungarian Method

The Hungarian method (developed by Hungarian mathematician D. Konig) of assignment provides us with an efficient method of finding the optimal solution without having to make a direct comparison of every solution. It works on the principle of reducing the given cost matrix to a matrix of opportunity costs.

Opportunity costs show the relative penalties associated with assigning resource to an activity as opposed to making the best or least cost assignment. If we can reduce the cost matrix to the extent of having at least one zero in each row and column, it will be possible to make optimal assignments (opportunity costs are all zero).

The Hungarian method (minimization case) can be summarized in the following steps:

Step 1. Develop the Cost Table From the Given Problem

If the number of rows are not equal to the number of columns and *vice versa*, a dummy row or dummy column must be added. The assignment costs for dummy cells are always zero.

Step 2. Find the Opportunity Cost Table

(a) Locate the smallest element in each row of the given cost table and then subtract that from each element of that row, and

(b) In the reduced matrix obtained from 2(a), locate the smallest element in each column and then subtract that from each element of that column. Each row and column now have at least one zero value.

Step 3. Make Assignments in the Opportunity Cost Matrix

The procedure of making assignments is as follows:

(a) Examine rows successively until a row with exactly one unmarked zero is obtained. Make an assignment to this single zero by making a brackets ( ) around it.

(b) For each zero value that becomes assigned, eliminate (strike off) all other zeros in the same row and/or column.

(c) Repeat Steps 3(a) and 3(b) for each column also with exactly single zero value cell that has not been assigned.

(d) If a row and/or column has two or more unmarked zeros and one cannot be chosen by inspection, then choose the assigned zero cell arbitrarily.

(e) Continue this process until all zeros in rows/columns are either enclosed (assigned) or struck off (×).

Step 4. Optimality Criterion

If the number of assigned cells is equal to the number of rows/columns, then it is an optimal solution. The total cost associated with this solution is obtained by adding original cost figures in the occupied cells.

If a zero cell was chosen arbitrarily in Step 3, there exists an alternative optimal solution. B optimal solution is found, then go to Step 5.

Step 5. Revise the Opportunity Cost Table

Draw a set of horizontal and vertical lines to cover all the zeros in the revised cost table obtained from Step 3, by using the following procedure:

(a) For each row in which no assignment was made, mark a tick (√)

(b) Examine the marked rows. If any zero cell occurs in those rows, mark a √ to the respective columns that contain those zeros.

(c) Examine marked columns. If any assigned zero occurs in those columns, tick the respective rows that contain those assigned zeros.

(d) Repeat this process until no more rows or columns can be marked.

(e) Draw a straight line through each marked column and each unmarked row.

If the number of lines drawn (or total assignments) is equal to the number of rows (or columns), the current solution is the optimal solution, otherwise go to Step 6.

Step 6. Develop the New Revised Opportunity Cost Table

(a) From among the cells not covered by any line, choose the smallest element. Call this value k.

(b) Subtract k from every element in the cell not covered by a line.

(c) Add k to every element in the cell covered by the two lines, *i.e.*, intersection of two lines.

(d) Elements in cells covered by one line remain unchanged.

SOLVED EXAMPLES

Example 1:

A department head has four tasks to be performed and three subordinates. The subordinates differ in efficiency. The estimates of the time, each subordinate would take to perform, are given below in the matrix. How should he allocate the tasks, one to each man, so as to minimize the total man hours?

Subordinates

		1	*2*	*3*
	I	*9*	*26*	*15*
Tasks	*II*	*13*	*27*	*6*
	III	*35*	*20*	*15*
	IV	*18*	*30*	*20*

Solution:

Since the matrix is not square, it is an unbalanced assignment problem. We introduce one fictitious subordinate (4th column with zero costs) to get a square matrix. Thus the resulting matrix is shown in the following table. Now the problem can be solved by usual method.

	1	2	3	4
I	9	26	15	0
II	13	27	6	0
III	35	20	15	0
IV	18	30	20	0

Step 1. Subtracting the minimum element of each row from every element of the corresponding row and then subtracting the minimum element of each column from every element of the corresponding column, the matrix reduces to

	1	2	3	4
I	0	6	9	0
II	4	7	0	0
III	26	0	9	0
IV	9	10	14	0

Step 2. Make the 'zero assignments' in usual manner. The illustration is shown in the following table.

	1	2	3	4
I	(0)	6	9	⊗
II	4	7	(0)	⊗
III	26	(0)	9	⊗
IV	9	10	14	(0)

Since in the table every row and every column have one assignment, so we have the complete optimal zero assignment as

I → 1, II → 3, III → 2.

Take IV will be left unassigned.

From the original matrix, the total time (man hours) = 9 + 6 + 20 = 35.

Example 2:

In a machine shop a supervisor wishes to assign five jobs among six machines any one of the job can be processed completely by any one of the machines the cost (in Rs.) of processing any jobs on any machine is given below.

Job

	A	*B*	*C*	*D*	*E*	*F*
1	*13*	*13*	*16*	*23*	*19*	*9*
2	*11*	*19*	*26*	*16*	*17*	*18*
3	*12*	*11*	*4*	*9*	*6*	*10*
4	*7*	*15*	*9*	*14*	*14*	*13*
5	*9*	*13*	*12*	*8*	*14*	*11*

The assignment of jobs to machines must be one-to-one basis. Assign the jobs to machines so that the total cost is minimum. Find the minimum total cost.

Solution:

We add a dummy sixth row (job 6) in the cost matrix so as to get the following balanced assignment problem:

	A	B	C	D	E	F
1	13	13	16	23	19	9
2	11	19	26	16	17	18
3	12	11	4	9	6	10
4	7	15	9	14	14	13
5	9	13	12	8	14	11
6	0	0	0	0	0	0

Now, following the usual procedure of solving an assignment problem, an optimum assignment is obtained as follows:

1 → F, 2 → A, 3 → E, 4 → C, 5 → D

with total minimum cost as

Rs. (9 + 11 + 6 + 9 + 8), *i.e.,* Rs. 43.

A company has 4 machines on which to do 3 jobs. Each job can be assigned to one and only one machine. The cost of each job on each machine is given in the following table:

	Machines			
Job	W	X	Y	Z
A	18	24	28	32
B	8	13	17	19
C	10	15	19	22

What are the job assignments which will minimize the cost?

A department head has six jobs and five subordinates. The subordinates differ in their efficiency and the tasks differ. In third intrinsic difficulty. The department head estimates the time each man would take to perform each task as given in the effectiveness matrix below:

Man				*Task*		
	A	B	C	D	E	F
1	20	15	26	40	32	12
2	15	32	46	26	28	20
3	11	15	2	12	6	14
4	8	24	12	22	22	20
5	12	20	18	10	22	15

Only one task can be assigned to one man determine how should the job be allocated so as to minimize the total man hour. Find the minimum total man hours.

Example 3(a):

A company has four territories open and four salesmen available for assignment. The territories are not equally rich in their sales potential; is estimated that a typical salesman operating in each territory would bring in the following annual sales:

Territory	:	I	II.	III	IV
Annual sales (Rs.)	:	60,000	50,000	40,000	30,000

Four salesmen are also considered to differ in their ability : it is estimated that, working under the same conditions, their yearly sales would be proportionately as follows:

Salesman	:	A	B	C	D
Proportion	:	7	5	5	4

If the criterion is maximum expected total sales, then intuitive answer is to assign the best salesman to the richest territory, the next best salesman to the second richest, and so on. Verify this answer by the assignment technique.

Solution:

To construct the effectiveness matrix.

The sum of proportions of sales of four salesmen

$$= 7 + 5 + 5 + 4 = 21.$$

If, we consider Rs 10,000 as one unit, then A's annual sales in the four territories are

$$\frac{7}{21}\times 6, \frac{7}{21}\times 5, \frac{7}{21}\times 4, \frac{7}{21}\times 3.$$

Similarly we can calculate the annual sales of other salesmen in different territories. In order to avoid fractional values we consider the sales in 21 years.

Thus, the maximum sale matrix is obtained as follows:

Sales in Rs 10,000 →		6	5	4	3
Sales proportion ↓		I	II	III	IV
7	A	42	35	28	21
5	B	30	25	20	15
5	C	30	25	20	15
4	D	24	20	16	12

This is a 'maximization' problem. To convert it into a 'minimization' one let us multiply each element of the above matrix by – 1. Thus resulting matrix becomes:

	I	II	III	IV
A	–42	–35	–28	–21
B	–30	–25	–20	–15
C	–30	–25	–20	–15
D	–24	–20	–16	–12

Now we shall solve the minimization problem by usual assignment algorithm.

Step 1. Subtracting the smallest element of each row from every element of that row and then subtracting the smallest element of each column from every element of that column the reduced matrix is shown in the following Table.

L_2 ↑				
(0)	3	6	9	
⊗	1	2	3	
⊗	1	2	3	
⊗	(0)	⊗	⊗	→ L_1

Step 2. Make zero assignments. Since total no. of marked '( )' zeros is less than 4, so an optimal assignment is not possible at this stage (as shown in the table of step 1).

Step 3. We draw minimum no. of lines to cover all the zeros.

The minimum element among all uncovered elements is 1.

Then subtract this value 1 from each uncovered element, add 1 to the elements which lie at the intersection of two lines L_1, L_2 and leave other elements unchanged. Thus, the revised matrix is shown in the table.

	L_2	L_3			
	(0)	2	5	8	√ (4)
	⊗	(0)	1	2	√ (5)
	⊗	⊗	1	2	√ (1)
L1	1	⊗	(0)	⊗	
	√ (2)	√ (3)			

Step 4. Giving the 'zero assignments' in the table of step 3, we see that there is no assignment in row 3 and column 4. So we again proceed to draw minimum number of lines to cover all zeros at least once.

Step 5. We find that three lines L_1, L_2, L_3 cover all the zeros. Again minimum element among all uncovered element is 1. Subtracting 1 from all the uncovered elements, adding 1 to the elements which lie at the intersection of the lines L_1, L_2, L_3 and leaving other elements unchanged, we get the following matrix:

0	2	4	7
0	0	0	1
0	0	0	1
2	1	0	0

Step 6. Giving 'zero assignments', we get two assignments as shown in the following Tables:

	I	II	III	IV
A	(0)	2	4	7
B	⊗	(0)	⊗	1
C	⊗	⊗	(0)	1
D	2	1	⊗	(0)

	I	II	III	IV
A	(0)	2	4	7
B	⊗	⊗	(0)	1
C	⊗	(0)	⊗	1
D	2	1	⊗	(0)

Thus, two optimal solution are

(i) A → I, B → II, C → III, D → IV

(i) A → I, B → III, C → II, D → IV.

From both the solution, it is obvious that the best salesman A is assigned to the richest territory I, the worst salesman D to the poorest territory IV. Salesman B and C being equally good, so they may be assigned to either II or III. This verifies the given intuitive answer.

Example 3(b):

A car hire company has one car at each of five depots a, b, c, d and e. A customer requires a car in each town namely A, B, C, D and E. Distance (in kms) between depots (origins) and towns (destinations) are given in the following distance matrix:

	a	*b*	*c*	*d*	*d*
A	*160*	*130*	*175*	*190*	*200*
B	*135*	*120*	*130*	*160*	*175*
C	*140*	*110*	*155*	*170*	*185*
D	*50*	*50*	*80*	*80*	*110*
E	*55*	*35*	*70*	*80*	*105*

How should cars be assigned to customers so as to minimize the distance travelled?

Solution:

Step 1. Subtracting the minimum element of each row form every element of the corresponding row, the matrix reduces to

30	0	45	60	70
15	0	10	40	55
30	0	45	60	75
0	0	30	30	60
20	0	35	45	70

Now, subtracting the minimum element of each column from every element of the corresponding column, the matrix reduces to

30	0	35	30	15
15	0	0	10	0
30	0	35	30	20
0	0	20	0	5
20	0	25	15	15

Step 2. Now, we give the zero assignments in our usual manner. Row 1 has a single zero in column 2. Make an assignment by marking '( )' around it and delete other zeros (if any) in column 2 by marking '×', Examining the set rows completely, an identical procedure is applied successively to columns.

30	0	35	30	15
15	⊗	0	10	⊗
30	⊗	35	30	20
0	⊗	20	⊗	5
20	⊗	25	15	15

Now column 1 has a single zero in row 4. Make an assignment by marking '( )' at this zero an cross the other zero of row 4 which in not yet crossed. Column 3 has a single zero in row 2, make an assignment at this zero by putting '( )' and cross the other zero of row which is not yet crossed. At this stage all zeros have been either assigned or crossed out. It is observed that row 3, 5, column 4 and column 5 each has no assignment. Hence, the required solution cannot be obtained at this stage.

Step 3. In this step we draw minimum number of lines to cover all zeros a least once. For this we proceed as follows:

(i) Mark (√) row 3 and row 5 as they have no assignments.

(ii) Mark (√) column 2 as having zeros in the marked rows 3 and 5.

(iii) Mark (√) row 1 as it contains assignment in the marked column 2.

Now further rows or colons will be required to mark during this procedure.

(iv) Now draw line L_1 through marked column 2. Then draw lines L_2 and L_3 through unmarked rows 2 and 4.

The required lines will be L_1, L_2 and L_3. No zero is left uncovered.

		L_1				
	30	(0)	35	30	15	√ 4
L_2	15	⊗	(0)	10	⊗	
	30	⊗	35	30	20	√ 1
L_3	(0)	⊗	20	⊗	5	
	20	⊗	25	15	15	√ 2
		√				
		3				

Step 4. In this step we select the smallest element among all uncovered elements of the matrix of step 3.

Here, this element is 15. Subtracting this element 15 from all the elements that do not have a line through them and adding to every element that lies at the intersecting of two lines and leaving the remaining elements unchanged we get, the following matrix.

15	0	20	15	0
15	15	0	10	0
15	0	20	15	5
0	15	20	0	5
5	0	10	0	0

Step 5. Now again performing the step 2 we make the zero assignments. It is observed that there are no remaining zeros and every row (column) has an assignment as shown in the table.

15	⊗	20	15	(0)
15	15	(0)	10	⊗
15	(0)	20	15	5
(0)	15	20	⊗	5
5	⊗	10	(0)	⊗

Thus the complete optimal assignment plan is given by

A → e, B → c, C → b, D → a, E → d.

Form the original matrix, the minimum cost (distance travelled)

= (200 + 130 + 110 + 50 + 80) kms. = 570 kms.

Example 3(c):

Solve assignment problem represented by the following matrix:

	I	*II*	*III*	*IV*	*V*	*VI*
A	*9*	*22*	*58*	*11*	*19*	*27*
B	*43*	*78*	*72*	*50*	*63*	*48*
C	*41*	*28*	*91*	*37*	*45*	*33*
D	*74*	*42*	*27*	*49*	*39*	*32*
E	*36*	*11*	*57*	*22*	*25*	*18*
F	*3*	*56*	*53*	*31*	*17*	*28*

Solution:

Step 1. Subtracting the minimum element of each row from every element of the corresponding row and then subtracting the minimum element of each column form every element of the corresponding column, the matrix reduces to:

0	13	49	0	0	13
0	35	29	5	10	0
13	0	63	7	7	0
47	15	0	20	2	0
25	0	46	9	4	2
0	53	50	26	4	20

Step 2. Make the 'zero assignments' in usual manner. The illustration is shown in the following table. Since row 3 and column 5 have no assignments so we proceed to the next step.

⊗	13	49	(0)	⊗	13
⊗	35	29	5	10	(0)
13	⊗	63	7	⊗	⊗
47	15	(0)	20	2	⊗
25	0	46	9	4	2
(0)	53	50	26	4	20

Step 3. Draw minimum number of lines to cover all zeros at least once. For this we proceed as follows:

(i) Mark (√) row 3 as having no assignment.

(ii) Mark (√) columns 2 and 6 as having zeros in marked row 3.

(iii) Mark (√) rows 5 and 2 as having assignments in the marked columns 2 and 6.

(iv) Mark (√) column 1 (not already marked) as having zero in the marked row 2.

(v) Then mark (√) row 6 as having assignment in the marked column 1.

⊗	13	49	(0)	⊗	13		
⊗	35	29	5	10	(0)	√	5
13	⊗	63	7	7	⊗	√	1
47	15	(0)	20	2	⊗		
25	(0)	46	9	4	2	√	4
(0)	53	50	26	4	20	√	7
√	√				√		
6	2				3		

Now draw lines L_1, L_2, L_3 through marked columns 1, 2, 6 respectively and L_4, L_5 through unmarked rows 1, 4 respectively. This way minimum set of five lines (5 < 6) to cover all the zeros is obtained.

Step 4. Now, the smallest element among all uncovered elements is 4. Subtracting this element 4 from all the uncovered elements, adding to every element that lies at the intersection of two lines and leaving the remaining elements unchanged the matrix of step 3 reduces to the new form as shown in the table.

4	17	49	0	0	17
0	35	25	1	6	0
13	0	59	3	3	0
51	19	0	20	2	4
25	0	42	5	0	2
0	53	46	22	0	20

Step 5. Repeating the step 2, make the 'zero assignments' as shown in the following table.

Thus exactly one marked '( )' zero in each row and each column of the matrix is obtained.

4	17	49	(0)	⊗	17
(0)	35	25	1	6	⊗
13	⊗	59	3	3	(0)
51	19	(0)	20	2	4
25	0	42	5	⊗	2
⊗	53	46	22	(0)	20

Thus the optimal assignment is

A → IV, B → I, C → VI,

D → III, E → II, F → V.

From the original matrix, minimum cost = 11 + 43 + 33 + 27 + 11 + 17 + 142.

Note: Another optimal solution this assignment problem is shown in the following table *i.e.,*

A → IV, B → VI, C → II, D → III, E → V, F → I.

4	17	49	(0)	⊗	17
⊗	35	25	1	6	(0)
13	(0)	59	3	3	⊗
51	19	(0)	20	2	4
25	⊗	42	5	(0)	2
(0)	53	46	22	⊗	20

From the original matrix, minimum cost

= 11 + 48 + 28 + 27 + 25 + 3 = 142.

Example 3(d):

Solve the minimal assignment problem whose effectiveness matrix is

	I	*II*	*III*	*IV*
A	*2*	*3*	*4*	*5*
B	*4*	*5*	*6*	*7*
C	*7*	*8*	*9*	*8*
D	*3*	*5*	*8*	*4*

Solution:

Step 1. Subtracting the minimum element of each row from every element of the corresponding row, the matrix reduces to

0	1	2	3
0	1	2	3
0	1	2	1
0	2	5	1

Now subtracting the minimum element of each column from every element of the corresponding column, the matrix reduces to

0	0	0	2
0	0	0	2
0	0	0	0
0	1	3	0

Step 2. Now test whether it is possible to make an assignment using only zeros. Here none of the rows or columns contain exactly one zeros, Therefore, we start with row 1 searching two zeros. While examining rows successively, it is observed that row 4 has two zeros. Now, arbitrarily make an assignment [indicated by '( )'] to one of these two zeros, say zeros say zero in the column 1 and cross other zeros in row 4 and column 1. Now we find column 1 which contains only one unmarked zero in row 3. We make assignment [indicated by '( )'] at this zero and cross all other zeros of this row. Now again we check the rows and columns for one unmarked zero. There is no such row or column. So we start with row 1 searching two unmarked zeros, say zero of column 2, and cross other zeros of row 1 (not already crossed) and column 2. Now the second row contains only one unmarked zero in third column where we can make an assignment [indicated by'( )'].

At this stage all zeros have been either assigned or crossed out. We observe that every row and every column have one assignment, so we have the complete 'zero assignment'.

Tables 1, 2, 3 show the necessary steps for reaching the optimal assignment.

Table 1

⊗	0	0	2
⊗	0	0	2
⊗	0	0	0
(0)	1	3	⊗

Table 2

⊗	0	0	2
⊗	0	0	⊗
⊗	⊗	⊗	(0)
(0)	1	3	⊗

Table 3

⊗	(0_	⊗	2
⊗	⊗	(0)	2
⊗	⊗	⊗	(0)
(0)	1	3	⊗

Thus we get the following optimal assignment ;

$$A \rightarrow II,\ B \rightarrow III,$$
$$C \rightarrow IV,\ D \rightarrow I.$$

Minimum cost = 3 + 6 = 8 = 3 = 20.

Note: In this example other optimal assignments are also possible. Students must try to find them. Each has the cost 20.

Example 4:

An airline that operates seven days a week, has the time table shown below. Crews must have a minimum layover of 5 hours between fights. Obtain the pair of flights that minimizes layover time away from home. For any given pair the crew will be based at the city that results in the smaller layover.

Delhi-Jaipur

Flight No.	*Departure*	*Arrival*
1	*7.00A.M.*	*8.00A.M.*
2	*8.00A.M.*	*9.00A.M.*
3	*1.30P.M.*	*2.30P.M.*
4	*6.30P.M.*	*7.30P.M.*

Jaipur-Delhi

Flight No.	*Departure*	*Arrival*
101	*8.00A.M.*	*9.15A.M.*
102	*8.30A.M.*	*9.45A.M.*
103	*12.00 Noon*	*1.15P.M.*
104	*5.30P.M.*	*6.45P.M.*

For each pair, mention the town where the crew should be based.

Solution:

First we construct the tables for layover times between the flights. Suppose we pair the flight no. 1 with flight no. 103 when crew is based at Delhi. Then the time of stay at Jaipur will be the layover time away from home. Now a plane of flight no. 1 which reaches Jaipur at 8.00 A.M., cannot fly at 12.00 Noon on the same day as minimum layover time is 5 hours. So it will depart Jaipur on the next day which will result in all layover time of 28 hours. Similarly, other layover times can be calculated.

Tables for layover times in hours

When crew based at Delhi

	101	102	103	104
1	24	24.5	28	9.5
2	23	23.5	27	8.5
3	17.5	18	21.5	27
4	12.5	13	16.5	22

When crew based at Jaipur

	101	102	103	104
1	21.75	21.25	17.75	12.25
2	22.75	22.25	18.75	13.25
3	28.25	27.75	24.25	18.75
4	9.25	8.75	5.25	23.75

To avoid the fractions we measure the layover times in terms of quarter hour (0.25 hr. or 15 minutes) as one unit of time. Thus multiplying the above tables by 4, the modified tables are as follows:

When crew based at Delhi

	101	102	103	104
1	96	98	112	38
2	92	94	108	34
3	70	72	86	108
4	50	52	66	88

When crew based at Jaipur

	101	102	103	104
1	87	85	71	49
2	91	89	75	53
3	113	111	97	75
4	37	35	21	95

As a next step we combine the above two tables, choosing that base which gives a lesser layover time for each pairing. The layover times marked with denote that crew is based at Jaipur, otherwise the crew is based at Delhi.

Minimum layover time table

	101	102	103	104
1	87*	85*	71*	38*
2	91*	89*	75*	34
3	70	72	86	75*
4	37*	35*	21*	88

Now, this is a usual minimal assignment problem. Solving it by usual assignment technique, finally we get the following Table.

	101	102	103	104
1	4*	0*	0*	0
2	12*	8*	8*	0
3	0	0	28	50*
4	4*	0*	0*	100

Giving the zero assignments, we the following Tables:

	101	102	103	104
1	4	0*	⊗	⊗
2	12*	8*	8*	0
3	0	⊗	28	50*
4	4*	⊗*	0*	100

	101	102	103	104
1	4*	⊗	0*	⊗
2	12*	8*	8*	0
3	0	⊗	28	50*
4	4*	0*	⊗*	100

From the above tables, two optimal assignments are

(i) (1 → 102)*, (2 → 104), (3 → 101), (4 → 103)*

(ii) (1 → 103)*, (2 → 104), (3 → 101), (4 → 102)*.

In both the cases minimum layover time is 210 quarter hours *i.e.,* 52 hours 30 minutes.

Example 5:

Four engineers are available to design four projects. Engineer 2 is not competent to design the project B. Given the following time estimates needed to each engineer to design a given project, find how should the engineers be assigned to projects so as to minimize the total design time of four projects.

Projects

		A	B	C	D
	1	12	10	10	8
	2	14	*Not suitable*	15	11
Engineers	3	6	10	16	4
	4	8	10	9	7

Solution:

To avoid the assignment 2 → B, we take its time to be very large. Then the cost matrix of the resulting assignment problem is shown in the following Table:

12	10	10	8
14	∞	15	11
6	10	16	4
8	10	9	7

Now, we apply the assignment technique in the usual manner. The following tables show the necessary steps for reaching the optimal solution:

4	2	2	0	3	(0)	⊗	⊗	3	(0)	⊗	1
3	∞	4	0	2	∞	2	(0)	1	∞	1	(0)
2	6	12	(0)	1	4	10	⊗	(0)	3	9	⊗
1	3	2	0	(0)	1	⊗	⊗	⊗	1	(0)	1

Thus, the optimal assignment is : 1 → B, 2 → D, 3 → A, 4 → C.

Total minimum cost (time) = 10 + 11 + 6 + 9 = 36.

Example 6:

A job shop has purchased 5 new machines of different types. Theorem 5 available locations in the shop where a machine could be installed. Some of these locations are more desirable than others for particular machines because of their proximity to work centres which would have a heavy work flow to and from these machines. Therefore, the objective is to assign the new machines to the available locations in order to minimize the total cost of material handling.

The estimated cost per unit time of materials handling involving each of the machines is given below for the respective locations. Locations 1, 2, 3, 4 and 5 are not considered suitable for machines A, B, C, D and E respectively. Find the optimal solution:

Location (Cost in Rs.)

Machines	*1*	*2*	*3*	*4*	*5*
A	*(×)*	*10*	*25*	*25*	*10*
B	*1*	*(×)*	*10*	*15*	*2*
C	*8*	*9*	*(×)*	*20*	*10*
D	*14*	*10*	*24*	*(×)*	*15*
E	*10*	*8*	*25*	*27*	*(×)*

How will the optimal solution get modified if location 5 is also unsuitable for machine A?

Solution:

Since locations 1, 2, 3, 4 and 5 are not suitable for machines A, B, C, D and E respectively, an extremely large cost (say ∞) should be attached to these locations. The resulting assignment problem cost matrix is shown below:

∞	10	25	25	10
1	∞	10	15	2
8	9	∞	20	10
14	10	24	∞	15
10	8	25	27	∞

∞	2	6	3	0
⊗	∞	0	2	1
⊗	3	∞	0	2
2	0	3	∞	3
⊗	0	6	5	∞

Now following the usual procedure of solving an assignment problem, an optimum assignment is obtained as displayed above. This gives us the optimum assignment as

$A \rightarrow 5$, $B \rightarrow 3$, $C \rightarrow 4$,

$D \rightarrow 2$, and $E \rightarrow 1$

with total minimum cost as Rs. (10 + 10 + 20 + 10 + 10), *i.e.*, Rs. 60.

Now, if location 5 is also unsuitable for machine A, we attach an extremely large cost (= ∞) to cell (1, 5). Now applying the assignment algorithm to this modified problem, the following assignment algorithm can easily be obtained:

$A \rightarrow 4$, $B \rightarrow 3$, $C \rightarrow 5$, $D \rightarrow 2$, and $E \rightarrow 1$

or $A \rightarrow 2$, $B \rightarrow 3$, $C \rightarrow 4$, $D \rightarrow 5$, and $E \rightarrow 1$

with minimum total cost of Rs. 65.

Example 7:

A department head has four subordinates, and four tasks have to be performed. Subordinates differ in efficiency and tasks differ in their intrinsic difficulty. Time each man would take to perform each task is given in the electiveness matrix below. How should the tasks be allocated, one to a man, so as to minimize the total man hours?

		Subordinates			
		I	*II*	*III*	*IV*
	A	*8*	*26*	*17*	*11*
Tasks	*B*	*13*	*28*	*4*	*26*
	C	*38*	*19*	*18*	*15*
	D	*19*	*26*	*24*	*10*

Solution:

We shall solve this problem step by step to understand the method described above.

Step 1. Subtracting the minimum element of each row from every element of the corresponding row, the matrix reduces to

	I	II	III	IV
A	0	18	9	3
B	9	24	0	22
C	23	4	3	0
D	9	16	14	0

Now, subtracting the minimum element of each column form every element of the corresponding column, the matrix reduces to

	I	II	III	IV
A	0	14	9	3
B	9	20	0	22
C	23	0	3	0
D	9	12	14	0

Step 2. Now test whether it is possible to make an assignment using only zeros.

Starting with row 1 of the matrix examine the rows one by one until a row containing exactly single zero element if found. We mark '( )' at this zero, *i.e.,* make an assignment and mark a cross '×' over all zeros if lying in the column continuing the assigned zero. Continue in this manner until all the rows have been examined. The illustration of the procedure is shown in he following table.

	I	II	III	IV
A	(0)	14	9	3
B	9	20	(0)	22
C	23	(0)	3	⊗
D	9	12	14	(0)

Now starting with column 1, examine all the columns until a column containing exactly one zero is found. We mark '( )' in that position *i.e.,* make an assignment cross other zeros if lying in the row containing this marked zero.

Continue in this manner until all the columns have been examined. The illustration of the procedure is shown in the following table.

	I	II	III	IV
A	(0)	14	9	3
B	9	20	(0)	22
C	23	(0)	3	⊗
D	9	12	14	(0)

At this stage all zeros have been either assigned or crossed out. We observe that every row and every column have one assignment, so we have the complete 'zero assignment' given by

$A \to I$, $B \to III$, $C \to II$, $D \to IV$.

This assignment is also optimal for the original matrix.

The minimum total man hours = 8 + 4 + 19 + 10 = 41.

Example 8:

A company has 5 jobs to be done. The following matrix shows the return in rupees on assigning ith (i = 1, 2, 3, 4, 5) machine to the jth job (j = A, B, C, D, E). Assign the five jobs to the five machines so as to maximize the total expected profit.

		Jobs				
		A	*B*	*C*	*D*	*E*
Machines	*1*	*5*	*11*	*10*	*12*	*4*
	2	*2*	*4*	*6*	*3*	*5*
	3	*3*	*12*	*5*	*14*	*6*
	4	*6*	*14*	*4*	*11*	*7*
	5	*7*	*9*	*8*	*12*	*5*

Solution:

First, we shall convect the problem from maximization to minimization. The greatest element of the given matrix is 14. Subtracting all the elements of the given matrix from 14, the modified matrix is obtained. Now we shall follow the usual procedure of solving an assignment problem.

9	3	4	2	10
12	10	8	11	9
11	2	9	0	8
8	0	10	3	7
7	5	6	2	9

Step 1. Subtracting the minimum element of each row from every element of the corresponding row and then subtracting the minimum element of each column from every element of the corresponding column, the modified matrix reduces to

3	1	2	0	7
0	2	0	3	0
7	2	9	0	7
4	0	10	3	6
1	3	4	0	6

Step 2. Now make 'zero assignments'. The following tables show the necessary steps for reaching the optimal assignment : 1 → C, 2 → E, 3 → D, 4 → B, 5 → A with maximum profit of Rs. 50.

3	1	2	(0)	7	2	⊗	1	⊗	6	1	⊗	(0)	⊗	5
(0)	2	⊗	3	⊗	⊗	2	(0)	4	⊗	⊗	3	⊗	5	(0)
7	2	9	⊗	7	6	1	8	(0)	6	5	1	7	(0)	5
4	(0)	10	3	6	4	(0)	10	4	6	3	(0)	9	4	5
1	3	4	⊗	6	(0)	2	3	⊗	5	(0)	3	3	1	5

Example 9:

Solve the travelling salesman problem given by the following data:

$$c_{12} = 20,\ c_{13} = 4,\ c_{14} = 10,\ c_{23} = 5,\ c_{34} = 6$$

$$c_{25} = 10,\ c_{35} = 6,\ c_{45} = 20,\ \text{where } c_{ij} = c_{ji}$$

and there is no route between cities i and j if the value for c_{ij} is not given above.

Solution:

Consider the given problem as an assignment problem.

Taking $c_{ij} = \infty$ for i = j, the cost matrix is as follows:

	1	2	3	4	5
1	∞	20	4	10	∞
2	20	∞	5	∞	10
3	4	5	∞	6	6
4	10	∞	6	∞	20
5	∞	10	6	20	∞

If there is no route between cities i and j then we have taken $c_{ij} = \infty$ to avoid the possibility of going from ith station to jth station.

Now we shall solve the problem by usual assignment algorithm. The following tables show the necessary steps for reaching the solution:

∞	15	(0)	4	∞
15	∞	⊗	∞	3
(0)	⊗	∞	⊗	⊗
4	∞	⊗	∞	12
∞	3	⊗	12	∞

∞	12	(0)	1	∞
12	∞	⊗	∞	(0)
(0)	⊗	∞	⊗	⊗
1	∞	⊗	∞	9
∞	(0)	⊗	9	∞

∞	12	(0)	⊗	∞
11	∞	⊗	∞	(0)
⊗	1	∞	(0)	1
(0)	∞	⊗	∞	9
∞	(0)	⊗	8	∞

Hence, the optimum solution of the assignment problem is:

$1 \rightarrow 3,\ 3 \rightarrow 4,\ 4 \rightarrow 1,\ 2 \rightarrow 5,\ 5 \rightarrow 2.$

But, this is not the solution to the travelling salesman problem, as it is not allowed to go from city 4 to 1 without visiting the cities 2 and 5.

We try to find the 'next best' solution which satisfies the additional restriction. The smallest element other than zero is 1. So we try to bring 1 into the solution. Since the element 1 occurs at two places, we shall consider both the cases separately until the acceptable solution is attained.

Make assignment in the cell (3, 2) having the element 1 instead of making zero assignment in the cell (5, 2). After making this assignment we observe that no other assignment can be made in the third row and second column. Consequently make assignment in the cell (5, 4) having element 8, instead of zero assignment. Consequently make assignment in the cell (5, 2).

The new assignment plan is shown in the following table:

		To				
		1	2	3	4	5
	1	∞	12	(0)	⊗	∞
	2	11	∞	⊗	∞	(0)
From	3	⊗	(1)	∞	⊗	1
	4	(0)	∞	⊗	∞	9
	5	∞	⊗	∞	(8)	∞

Thus, the resulting feasible solution is

$1 \rightarrow 3 \rightarrow 2 \rightarrow 5 \rightarrow 4 \rightarrow 1$ with cost 9.

Again if we make assignment in the cell (3, 5) having the next best element 1 instead of 0 marked in cell (3, 4), then no feasible solution is obtained in terms of zeros or with cost less than 9.

Hence the optimum route is

$$1 \rightarrow 3 \rightarrow 2 \rightarrow 5 \rightarrow 4 \rightarrow 1$$

The corresponding cost

$$= 4 + 5 + 10 + 20 + 10 = 49.$$

Example 10:

Given the matrix of setup costs, show how to sequence the production so as to minimize the setup cost per cycle.

		To				
		A_1	A_2	A_3	A_4	A_5
	A_1	∞	2	5	7	1
	A_2	6	∞	3	8	2
From	A_3	8	7	∞	4	7
	A_4	12	4	6	∞	5
	A_5	1	3	2	8	∞

Solution:

Consider the problem as an assignment. Applying the assignment technique, we get the following matrix, showing a solution in terms of marked '( )' zero:

		To				
	A_1	A_2	A_3	A_4	A_5	
	A_1	∞	1	3	6	(0)
From	A_2	4	∞	(0)	6	⊗
	A_3	4	3	∞	(0)	3
	A_4	8	(0)	1	∞	1
	A_5	(0)	2	⊗	7	∞

The solution to the assignment problem given by above matrix is

$A_1 \rightarrow A_2$, $A_5 \rightarrow A_1$,

$A_2 \rightarrow A_3$, $A_3 \rightarrow A_4$,

$A_4 \rightarrow A_2$.

This solution indicates to produce the products A_1, then A_5 and then again A_1, without producing the products A_2, A_3 and A_4, which violates the additional restriction of prodding each product once and only once before returning to the first product. So this is not a solution of the travelling salesman problem.

Now, we try to find the next best solution which also satisfies the additional restriction. The next minimum element (non-zero) in the matrix is 1. We try to bring 1 in the solution. The cost 1 also occurs at three places.

Start by making unity-assignment in the cell (1, 2) instead of zero assignment in the cell (1, 5). Then no other assignment can be made in the first row and the second column. The best solution of the problem lies in the marked '( )' elements as shown in the table.

		A_1	A_2	A_3	A_4	A_5
	A_1	∞	(1)	3	6	⊗
	A_2	4	∞	(0)	6	⊗
From	A_3	4	3	∞	(0)	3
	A_4	8	⊗	1	∞	(1)
	A_5	(0)	2	⊗	7	∞

Thus, the required solution of the problem is

$$A_1 \to A_2 \to A_3 \to A_4 \to A_5 \to A_1.$$

For this solution, the cost in the reduced matrix is 2.

On the other hand if we select the element 1 in the cell (4, 3) in the solution, then no feasible solution is available in terms of zero or for which the reduced matrix gives the minimum cost less than 2.

Hence, the most suitable sequence is

$$A_1 \to A_2 \to A_3 \to A_4 \to A_5 \to A_1.$$

The minimum setup cost = 2 + 3 + 4 + 5 + 1 = 15.

Example 11(a):

A salesman estimate that the following would be the cost on his route visiting the six cities as shown.

				To city			
		1	*2*	*3*	*4*	*5*	*6*
	1	∞	*20*	*23*	*27*	*29*	*24*
	2	*21*	∞	*19*	*26*	*31*	*24*
From city	*3*	*26*	*28*	∞	*15*	*36*	*26*
	4	*25*	*16*	*25*	∞	*23*	*18*
	5	*23*	*20*	*23*	*31*	∞	*10*
	6	*27*	*18*	*12*	*35*	*16*	∞

The sales man can visit each of the cities once and only once. Determine the optimum sequence he should follow to minimize the total distance travelled. What as the total distance travelled?

Solution:

Following the usual procedure of assignment algorithms, we obtain Table showing an optimum solution indicated by encircled zeros.

This table provides the optimum assignment schedule

$1 \rightarrow 3 \rightarrow 4 \rightarrow 2 \rightarrow 1$ and $5 \rightarrow 6 \rightarrow 5$

with distance according to this optimum route being 101.

	1	2	3	4	5	6
1	∞	0	0	7	2	14
2	0	∞	0	10	8	8
3	6	13	∞	0	14	11
4	4	0	6	∞	0	2
5	8	30	10	21	∞	0
6	13	6	0	26	0	∞

This table does not provide the solution to the travelling salesman problem, as it gives $1 \rightarrow 3$, $3 \rightarrow 4$, $4 \rightarrow 2$, $2 \rightarrow 1$, while city 2 is not allowed to follow city 1 unless city 5 and city 6 processed.

Now, we try to find the next best solution which also satisfies this extra restriction. The next minimum (non-zero) element in the matrix is 2. Therefore, we try to bring 2 into the solution. But the element 2 occurs at two places. We shall consider both the cases separately until the acceptable solution is attained,

	1	2	3	4	5	6
1	∞	0	0	7	(2)	14
2	0	∞	0	10	8	8
3	6	13	∞	0	14	11
4	4	0	6	∞	0	2
5	8	30	10	21	∞	0
6	13	6	0	26	0	∞

We start with the element (1, 5) and make the assignment in this cell instead of zero assignment in the cell (1, 3). After making this assignment we see that no other assignment can be made in the first row and third column, and thus the resulting feasible solution will be

$1 \to 5,\ 5 \to 6,\ 6 \to 3,\ 3 \to 4,\ 4 \to 2,\ 2 \to 1.$

The selected elements for this solution are encircled in Table . The cost corresponding to this feasible solution is 2.

Again, if we make assignment in the cell (4, 6) instead of (4, 2) then no feasible solution is available in terms of zeros. Hence, the best programme is

$$1 \to 5 \to 6 \to 3 \to 4 \to 2 \to 1.$$

The total set-up cost according to this route comes out to be 103.

Solve the travelling salesman problem given by the following data:

$$c_{12} = 20,\ c_{13} = 4,\ c_{14} = 10,$$
$$c_{23} = 5,\ c_{24} = 6,\ c_{25} = 10,$$
$$c_{35} = 6 \text{ and } c_{45} = 20,$$

where $c_{ij} = c_{ij}$

and there is no route between cities i and j if a value for c_{ij} is not shown above.

Given the following matrix of set-up cost, show how to sequence production so as to minimize set-up cost per cycle:

		To				
		A	B	C	D	E
	A	∞	2	5	7	1
	B	6	∞	3	8	2
	C	8	7	∞	4	7
From	D	12	4	6	∞	5
	E	1	3	2	8	∞

A medical representative has to visit five stations A, B, C, D, and E. He does not want to visit any station twice before completing his four of all the stations, and wishes to return to the starting station. Costs of going from one stations to another are given below. Determine the optimal route:

	A	B	C	D	E
A	∞	2	4	7	1
B	5	∞	2	8	2
C	7	6	∞	4	6
D	10	3	5	∞	4
E	1	2	2	8	∞

Example 11(b):

A company is faced with the problem of assigning six different machines to five different jobs. The costs are estimated as follows (in hundreds of rupees):

		Jobs				
		1	*2*	*3*	*4*	*5*
	1	*2.5*	*5*	*1*	*6*	*1*
	2	*2*	*5*	*1.5*	*7*	*3*
Machines	*3*	*3*	*6.5*	*2*	*8*	*3*
	4	*3.5*	*7*	*2*	*9*	*4.5*
	5	*4*	*7*	*3*	*9*	*6*
	6	*6*	*9*	*5*	*10*	*6*

Solve the problem assuming that the objective is to minimize the total cost.

Solution:

Since the matrix is not square, it is an unbalanced assignment problem. We introduce one fictitious job (6th column with zero costs) to get a square matrix. Further the matrix invoives elements is decimal. We can make then complete numbers 'by multiplying each element of the cost matrix by 2. The new modified cost matrix is shown in the following Table.

	1	2	3	4	5	6
1	5	10	2	12	2	0
2	4	10	3	14	6	0
3	6	13	4	16	6	0
4	7	14	4	18	9	0
5	8	14	6	18	12	0
6	12	18	10	20	12	0

Applying the usual procedure, finally we get the following Table:

	1	2	3	4	5	6
1	2	0	0	0	0	4
2	0	0	1	2	2	4
3	0	1	0	2	0	2
4	1	2	0	4	3	2
5	0	0	0	2	4	0
6	4	4	4	4	4	0

Giving the zero assignments, we get the following tables:

	1	2	3	4	5	6
1	2	⊗	⊗	(0)	⊗	4
2	(0)	⊗	1	2	2	4
3	⊗	1	⊗	2	(0)	2
4	1	2	(0)	4	3	2
5	⊗	(0)	⊗	2	4	⊗
6	4	4	4	4	4	(0)

	1	2	3	4	5	6
1	2	⊗	⊗	(0)	⊗	4
2	⊗	(0)	1	2	2	4
3	⊗	1	⊗	2	(0)	2
4	1	2	(0)	4	3	2
5	(0)	⊗	⊗	2	4	⊗
6	4	4	4	4	4	(0)

Thus the original problem has two alternative optimum solution:

(i) $1 \to 4, 2 \to 1, 3 \to 5, 4 \to 3, 5 \to 2$

(ii) $1 \to 4, 2 \to 2, 3 \to 5, 4 \to 3, 5 \to 1$.

The 6th machine will be left unassigned.

From the original matrix, the total minimum cost = 20 *i.e.*, Rs. 2000 (in each case).

EXERCISES

1. The owner of a small machine shop has four persons variable to assign to jobs for the day. Five jobs are offered with the expected profit in rupees for each person on each job being as follows:

	Job A	B	C	D	E
Persons 1	6.20	7.80	5.00	10.10	8.20
2	7.10	8.40	6.10	7.30	5.90
3	8.70	9.20	11.10	7.10	8.10
4	4.80	6.40	8.70	7.70	8.00

Find the assignment of persons to jobs that will result in a maximum profit. Which job shadow declined?

2. Give an algorithm to solve an assignment problem.

3. How can the travelling select man problem be solved using assignment algorithm.

4. There are 3 persons P_1, P_2 and P_3 and 5 jobs $J_1, J_2, \ldots J_5$. Each person can do only one job and a job is to be done by one person only. Using

Hungarian method, find which 2 jobs should be left undone in the following cost minimizing assignment problem.

	J_1	J_2	J_3	J_4	J_5
P_1	7	8	6	5	9
P_2	9	6	7	6	10
P_3	8	7	9	5	6

5 (a) Give an algorithm to solve an assignment problem.

(b) Show that an assignment problem is a special case of a transportation problem.

6. Explain how could an assignment problem be solved using the transportation approach?

7. Use the Hungarian method to find which of the two jobs should be left undone when each of the 4 persons will do only one job in the following cost minimizing assignment problem:

		Job					
		J_1	J_2	J_3	J_4	J_5	J_6
	P_1	10	9	11	12	8	5
	P_2	12	10	9	11	9	4
Persons	P_3	8	11	10	7	12	4
	P_4	10	7	8	10	10	5

8. The jobs A, B, C are to be assigned to three machines X, Y, Z. The processing costs (Rs.) are as given in the matrix shown below. Find the allocation which will minimize the overall processing cost.

		Machines		
		X	Y	Z
	A	19	28	31
Job	B	11	17	16
	C	12	15	13

9. Use Hungarian method to solve the following cost minimizing assignment problem:

10. A computer centre has got three expert programmers. The center needs three application programmed to be developed. The head of the

computer centre, after studying carefully the programmed to be developed. Estimates the computer time in minutes required by the experts to the application programmes as follows:

		Programmes		
		A	*B*	*C*
	1	120	100	80
Programmers	2	70	90	110
	3	110	140	120

Assign the programmers to the programmes in such a way that total computer time is least.

11. Find the optimal solution for the assignment problem with the following cost matrix:

	I	II	III	IV	V
A	11	17	8	16	20
B	9	7	12	6	15
C	13	16	15	12	16
D	21	24	17	28	26
E	14	10	12	11	15

12. Give the mathematical formulation of an assignment problem. How does it differ from a transportation problem?

13. State and discuss the methods for solving an assignment problem. How is the Hungarian method better than other methods for solving an assignment problem?

14. Five men are available to do five different jobs. From past records, the time (in hours) that each man takes to do each job is known and given in the following table:

				Jobs		
		I	*II*	*III*	*IV*	*V*
	A	2	9	7	7	1
	B	6	8	7	6	1
Men	C	4	6	5	3	1
	D	4	2	7	3	1
	E	5	3	9	5	1

Find the assignment of men to jobs that will minimize the total time taken.

15. A pharmaceutical company is producing a single product and is selling it through five agencies situated in different cities. All of a sudden, there is a demand for the product in another five cities not having any agency of the company. The company is faced with the problem of deciding on how to assign the existing agencies to despatch the product to needy cities in such a way that the travelling distance is minimized. The distance between the surplus and deficit cities (in km) is given in the following table.

Deficit cities

		a	*b*	*c*	*d*	*e*
	A	160	130	115	190	200
	B	135	120	130	160	175
Surplus cities	C	140	110	125	170	185
	D	50	50	80	80	110
	E	55	35	80	80	105

Determine the optimum assignment schedule.

16. A national truck rental service has a surplus of one truck in each of the cities, 1, 2, 3, 4, 5 and 6; and a deficit of one truck in each of the cities 7, 8, 9, 10, 11 and 12. The distances (in km) between the cities with a surplus and cities with deficit are displayed in the table:

To

	7	*8*	*9*	*10*	*11*	*12*
1	31	62	29	42	15	41
2	12	19	39	55	71	40
3	17	29	50	41	22	22
From 4	35	40	38	42	27	33
5	19	30	29	16	20	23'
6	72	30	30	50	41	20

How should the trucks be displayed so as to minimize the total distance travelled?

17. An air freight company picks up and delivers freight where customers want. The company has two types of aircraft, X and Y, with equal loading capacities but different operations costs. These are shown below:

Type of Aircraft	*Operating Costs (Rs)*	
	Empty	*Loaded*
X	1.00	2.00
Y	1.50	3.00

The present four locations of the aircrafts which the company has are; J – X; K – Y; L – Y, and M – X. Four customers of the company located at A, B, C and D want to transport nearly the same size of load to their final destinations. The final destinations are 600, 300, 1,000 and 500 km, from the loading points A, B, C and D, respectively. Distances (in km) between the aircraft and the loading points are as follows:

		Loading Point			
		A	***B***	***C***	***D***
	J	200	200	400	100
Aircraft	K	300	100	300	300
Location	L	400	100	100	500
	M	200	200	400	200

Determine the allocations which minimize the total cost of transportation.

18. A departmental head has four subordinates and four tasks to be performed. The subordinates differ in efficiency and the tasks differ in their intrinsic difficulty. His estimates of the times that each man would take to perform each task is given below in the matrix:

		Tasks			
		I	II	III	IV
Subordinates	A	8	26	17	11
	B	13	28	4	26
	C	38	19	18	15
	D	19	26	24	10

How should the tasks be allocated to subordinates so as to minimize the total man-hours?

19. State the travelling salesman problem and formulate it as an assignment problem.

20. An automobile dealer wishes to put four repairmen to four different jobs. The repairmen have somewhat different kinds of skills and they exhibit different levels of efficiency from one job to another. The

dealer has estimated the number of man-hours that would be required for each job-man combination. This is given in matrix form in the following table:

		Jobs			
		A	***B***	***C***	***D***
	1	5	3	2	8
	2	7	9	2	6
Men	3	6	4	5	7
	4	5	7	7	8

Find the optimal assignment that will result in minimum man-hours needed.

21. Explain the conceptual justification that an assignment problem can be viewed as a linear programming problem.
22. Specify the dual of an assignment problem. What are the techniques used in solving an assignment problem?
23. The personnel manager of ABC Company wants to assign Mr. X, Mr. Y and Mr Z to regional offices. But the firm also has an opening in its Chennai office and would send one of the three to that branch if it were more economical than a move to Delhi, Mumbai or Kolkata. It will cost Rs. 2,000 to relocate Mr. X to Chennai, Rs. 1,600 to reallocate Mr. Y there, and Rs. 3,000 to move Mr. Z. What is the optimal assignment of personnel to offices?

		Office		
		Delhi	***Mumbai***	***Kolkata***
	Mr X	1,600	2,200	2,400
Personnel	Mr Y	1,000	3,200	2,600
	Mr Z	1,000	2,000	4,600

24. Can there be multiple optimal solutions to an assignment problem? How would you identify the existence of multiple solutions, if any?
25. How would you deal with the assignment problems, where (a) the objective function is to be maximized? (b) some assignments are prohibited?
26. Explain how to modify an effectiveness matrix in an assignment problem if a particular assignment is prohibited.
27. What is an unbalanced assignment problem? How is the Hungarian method applied for obtaining a solution if the matrix is rectangular?

28. A project work consists of four major jobs for which an equal number of contractors have submitted tenders. The tender amount quoted (in lakhs of rupees) is given in the matrix.

		Job			
		a	*b*	*c*	*d*
	1	10	24	30	15
Contractor	2	16	22	28	12
	3	12	20	32	10
	4	9	26	34	16

Find the assignment which minimizes the total cost of the project, when each contractor has to be assigned at least one job.

29. A company has four zones open and four salesmen available for an assignment. The zones are not equally rich in their sales potential. It is estimated that a typical salesman operating in each zone would bring in the following annual sales:

Zone	:	A	B	C	D
Annual Sates (Rs)	:	60,000	50,000	40,000	30,000

The four salesmen also differ in their ability. It is estimated that, working under the same conditions, their yearly sales would be proportionately as follows:

Salesmen	:	P	Q	R	S
Proportion	:	7	5	5	4

If the criterion is maximum expected total sales, the best answer is to assign the best salesman to the richest zone, the next to the second richest zone and so on. Verify this by the method of assignment.

30. A company has four machines which are to be used for three jobs. Each job can be assigned to one and only one machine. The cost of each job on each machine is given in the following table.

		Machines			
		W	*X*	*Y*	*Z*
	A	18	24	28	32
Jobs	B	8	13	17	18
	C	10	15	19	22

What are the job assignment pairs which shall minimize the cost?

31. Five workers are available to work with the machines and the respective costs (in rupees) associated with each worker machine assignment are

given below. A sixth machine is available to replace one of the existing ones and the associated costs are also given below.

Machines

		M_1	M_2	M_3	M_4	M_5	M_6
	W_1	12	3	6	–	5	9
	W_2	4	11	–	5	–	8
Workers	W_3	8	2	10	9	7	5
	W_4	–	7	8	6	12	10
	W_5	5	8	9	4	6	1

(a) Determine whether the new machine can be accepted.

(b) Also determine optimal assignment and the associated saving in cost.

32. If (x_{ij}), $i = 1, 2, \ldots, n$; $j = 1, 2, \ldots, n$ is an optimal solution for an assignment problem with cost (c_{ij}), then it is also optimal for the problem with cost (c_{ij}') when

$c_{ij}' = c_{ij}$ for $i, j = 1, 2, \ldots, n; j \neq k$

$c_{ik}' = c_{ik} - A$, where A is a constant.

33. A firm is contemplating the introduction of three products 1, 2 and 3, in its three plants A, B and C. Only a single product is decided to be introduced in each of the plants. The unit cost of producing a product in a plant, is given in the following matrix.

Plant

		A	*B*	*C*
	1	8	12	–
Product	2	10	6	4
	3	7	6	6

(a) How should the product be assigned so that the total unit cost is minimized?

(b) If the quantity of different products to be produced is as follows, then what assignment shall minimize the aggregate production cost?

Product	***Quantity (in units)***
1	2,000
2	2,000
3	10,000

(c) What would your answer be if the three products were to be produced in equal quantities?

(d) It is expected that the selling prices of the products produced by different plants would be different, as shown in the following table:

		Plant		
		A	*B*	*C*
Product	1	15	18	–
	2	18	16	10
	3	12	10	8

Assuming the quantities mentioned in (b) above would be produced and sold, how should the products be assigned to the plants to obtain maximum profits?

34. A fast-food chain wants to build four stores. In the past, the chain has used six different construction companies, and having been satisfied with each, has invited each to bid on each job. The final bids (in lakhs of rupees) were as shown in the following table:

	Construction Companies					
	1	*2*	*3*	*3*	*4*	*5*
Store 1	85.3	88.0	87.5	82.4	89.1	86.7
Store 2	78.9	77.4	77.4	76.5	79.3	78.3
Store 3	82.0	81.3	82.4	80.6	83.5	81.7
Store 4	84.3	84.6	86.2	83.3	84.4	85.5

Since the fast-food chain wants to have each of the new stores ready as quickly as possible, it will award at the most one job to a construction company. What assignment results in minimum total cost to the fast-food chain?

35. A methods engineer wants to assign four new methods to three work centres. The assignment of the new methods will increase production and they are given below. If only one method can be assigned to a work centre, determine the optimum assignment:

		Increase in production (unit) Work Centres		
		A	*B*	*C*
Method	1	10	7	8
	2	8	9	7
	3	7	12	6
	4	10	10	8

36. Consider a problem of assigning four clerks to four tasks. The time (hours) required to complete the task is given below:

		Tasks			
		A	*B*	*C*	*D*
	1	4	7	5	6
Clerks	2	–	8	7	4
	3	3	–	5	3
	4	6	6	4	2

Clerk 2 cannot be assigned to task A and clerk 3 cannot be assigned to task B. Find all the optimum assignment schedules.

37. Explain the difference between a transportation problem and an assignment problem.

38. The marketing director of a multi-unit company is faced with a problem of assigning 5 senior managers to six zones.

From past experience he knows that the efficiency percentage judged by sales, operating costs, etc., depends on manager-zone combination. The efficiency of different managers is given below:

		Zones					
		I	*II*	*III*	*IV*	*V*	*VI*
	A	73	91	87	82	78	80
	B	81	85	69	76	74	85
Manager	C	75	72	83	84	78	91
	D	93	96	86	91	83	82
	E	90	91	79	89	69	76

Find out which zone will be managed by a junior manager due to non-availability of a senior manager.

39. A head of department in a college has the problem of assigning courses to teachers with a view to maximizing educational quality in his department. He has available to him one professor, two associate professors, and one teaching assistant (TA).

Four courses must be offered and, after appropriate evaluation, he has arrrived at the following relative ratings (100 = best rating) regarding the ability of each instructor to teach each of the four courses.

		Course 1	*Course 2*	*Course 3*	*Course 4*
Prof.	1	60	40	60	70
Prof.	2	20	60	50	70
Prof.	3	20	30	40	60
TA		30	10	20	40

How should he assign his staff to the courses to realize his objective?

40. At the end of a cycle of schedules, a transport company has a surplus of one truck in +ach of the cities 1, 2, 3, 4, 5 and a deficit of one truck in each of the cities A, B, C, D, E and F. The distances (in kilometres) between the cities with a surplus and cities with a deficit are given below:

To City

		A	*B*	*C*	*D*	*E*	*F*
	1	80	140	80	100	56	98
	2	48	64	94	126	170	100
From City	3	56	80	120	100	70	64
	4	99	100	100	104	80	90
	5	64	80	90	60	60	70

How should the trucks be despatched so as to minimize the total distance travelled? Which city will not receive a truck?

41. A company is considering an expansion into five new sales territories. The company has recruited four new salesmen. Based on the salesmen's experience and personality traits, the sales manager has assigned a rating to each of the salesmen for each of the sales territories. The ratings are as follows:

Territory

		1	*2*	*3*	*4*	*5*
	A	75	80	85	70	90
	B	91	71	82	75	85
Salesmen	C	78	90	85	80	80
	D	65	75	88	85	90

Suggest optimal assignment of the salesmen. If for certain reasons, salesman D cannot be assigned to territory 3, will the optimal assignment be different? If so, what would be the new assignment schedule?

42. Solve the travelling salesman problem in the matrix shown below:

	1	2	3	4	5
1	∞	6	12	6	4
2	6	∞	10	5	4
3	8	7	∞	11	3
4	5	4	11	∞	5
5	5	2	7	8	∞

43. The personnel manager of a medium-sized company decides to recruit two employees D and E in a particular section of the organization. The section has five fairly defined tasks 1, 2, 3, 4 and 5; and three employees A, B and C are already employed in the section. Considering the specialized nature of task 3 and the special qualifications of the recruit D for task 3, the manager decides to assign task 3 to employee D and then assign the remaining tasks to remaining employees so as to maximize the total effectiveness. The index of effectiveness of each employee of different tasks is as under.

	Tasks				
	1	*2*	*3*	*4*	*5*
A	25	55	60	45	30
Employees B	45	65	55	35	40
C	10	35	45	55	65
D	40	30	70	40	60
E	55	45	40	55	10

Assign the tasks for maximizing total effectiveness. Critically examine whether the decision of the manager to assign task 3 to employee D was correct.

44. Solve the 'travelling salesman problem' given by the following data:

$c_{12} = 4$, $c_{13} = 7$, $c_{14} = 3$, $c_{23} = 6$, $c_{24} = 3$ and $c_{34} = 7$, where $c_{ij} = c_{ji}$.

45. Describe the Hungarian method of solving the assignment problem.

46. Five swimmers are eligible to compete in a relay team which is to consist of four swimmers swimming four different swimming styles; back stroke, breast stroke, free style and butterfly. The time taken by the five swimmers—Anand, Bhasker, Chandru, Dorai and Easwar to cover a distance of 100 metres in various swimming styles is given below in minutes, seconds.

Anand swims the back stroke in 1 : 09, the breast stroke in 1 : 15 and has never competed in the free style or butterfly.

Bhasker is a free style specialist averaging 1 : 01 for the 100 metres but can also swim. the breast stroke in 1 : 16 and butterfly in 1 : 20.

Chandru swims all styles: back 1:10, butterfly 1:12, free style 1 : 05 and breast stroke 1 : 20.

Dorai swims only the butterfly 1 : 11, while Easwar swims the back stroke 1 : 20, breast stroke 1 : 16, free style 1 : 06 and the butterfly 1 : 10.

Which swimmer should be assigned to which swimming style? Who will not be in the relay?

47. Products 1, 2, 3, 4 and 5 are to be processed on a machine. The set-up costs in rupees per change depend upon the product presently on the machine and the set-up to be made and are given by the following data:

    $C_{12} = 16$, $C_{13} = 4$, $C_{14} = 12$, $C_{23} = 6$, $C_{24} = 5$, $C_{25} = 8$, $C_{35} = 6$, $C_{45} = 20$; $C_{ij} = C_{ji}$, $C_{ij} = \infty$ for t = j for all values of / andj not given in the data. Find the optimum sequence of products in order to minimize the total set-up cost.

48. Write a short note on travelling salesman problem.

49. Five wagons are available at five stations 1, 2, 3, 4, and 5. These are required at five stations I, II, III, IV and V.

50. What is an assignment problem? Give two areas of its applications.

51. Give the mathematical formulation of an assignment problem.

4

Sensitivity Analysis

INTRODUCTION

The process of studying the sensitivity of the optimal solution of an LP problem is also called *post-optimality analysis* because it is done after an optimal solution, assuming a given set of parameters, has been obtained for the model. Sensitivity analysis is the study of sensitivity of the optimal solution of an LP problem due to discrete variations (changes) in its parameters.

The degree of sensitivity of the solution due to these variations can range from no change at all to a substantial change in the optimal solution of the given LP problem. Thus, in sensitivity analysis, we determine the range over which the LP model parameters can change without affecting the current optimal solution.

For this, instead of resolving the entire problem as a new problem with new parameters, we may consider the original optimal solution as an initial solution for the purpose of knowing the ranges, both lower and upper, within which a parameter may assume a value.

Different categories of parameter changes in the original LP model, discussed in this chapter include:

(i) Profit (or cost) per unit (c_j) associated with both basic and non-basic decision variables (coefficients in the objective function).

(ii) Availability of resources (right-hand side of constants, b_i).

(iii) Consumption of resources per unit of decision variables x_j (coefficients of decision variables on the left-hand side of constraints, a_{ij}).

(iv) Addition of a new variable to the existing list of variables in LP problem.

(v) Addition of a new constraint to the original LP problem constraints.

CHANGE IN OBJECTIVE FUNCTION COEFFICIENT (c_j)

Suppose the coefficient c, in the objective function of an LP model represents either profit or cost per unit of an activity (variable) x_j. Then the question that may arise is: *What happens to the optimal solution and the objective function value when this coefficient is changed*? For example, let Rs. 10 be the per unit profit coefficient of a particular variable in the objective function.

After obtaining an optimal solution to the LP problem with Rs. 10 as one of the objective function coefficients, the decision-maker realises that its true value might be any value between Rs. 9 and Rs. 11 per unit. The test of sensitivity of the objective function value with respect to this coefficient (or on any other such coefficients) determines the range (both lower and upper) of values within which each c_j ($j = 1, 2, ..., n$) can lie without changing the current optimal solution. Such an analysis can help the decision-maker in deciding whether resources from other activities (variables) should be diverted to (diverted away from) a more profitable (or less profitable) activity. Changes in the profit or cost coefficients (contributions) in the objective function can occur for a *basic variable* (the variable in the solution mix) or a non-basic variable (the variable not in the solution mix). The sensitivity ranges for these variables are determined differently. Thus, these two cases will be discussed separately. Given an optimal basic feasible solution [$x_B = B^{-1}b$] with basis matrix B, suppose that the coefficient c_k of a variable x_k is changed from c_k to $c_k + \Delta c_k$.

Since the optimal basic feasible solution or solution values appeared in the 'x_B' column of the simplex table do not involve cost (or profit) coefficients, c_j in their calculations, it will remain feasible even for any change in the coefficients in the objective function. However, such a change in coefficients (c_j's) may affect the optimality of this solution. In other words, as $c_k - z_k$ ($= c_k - c_B B^{-1} a_k$) involves coefficients c_k, the effect of this change will be seen in the $c_j - z_j$ row of the optimal simplex table.

Case I : Change in the coefficient of a non-basic variable

The current optimal solution for a maximization LP problem will remain optimal as long as all $c_j - z_j \leq 0$, for all j. Let c_k be the coefficient of a non-basic variable x_k in the objective function. Since c_k is the coefficient of non-basic variable x_k, therefore, it does not affect any of the c_j values listed in the 'c_B' column of simplex table, associated with basic variables. Since the calculation of $z_j = c_B B^{-1} a_j$ values do not, involve c_j, therefore changes in c_j does not alter z_j values and hence ($c_j - z_j$) values remain unchanged except $c_k - z_k$ value due to change in c_k. In other words, any change in this coefficient does not affect feasibility of the optimal solution. This means that unit profit of x_k can be lowered to any level without causing the optimal solution to change. But any increase in its unit profit beyond a certain level (i.e. upper limit) should make this variable eligible to be a basic variable in the new solution mix. Obviously, then $c_k - z_k$ will no longer be negative.

To retain optimality of the current optimal solution for a change Δc_k in c_k, we must have $(c_k + \Delta c_k) - z_k \leq 0$ or $c_k + \Delta c_k \leq z_k$. Hence, for an LP problem with an objective function of maximization type, the value of c_k may be increased up to the value of z_k, and decrease to negative infinity $(-\infty)$ without affecting the optimal solution.

Case II : Change in the coefficient of a basic variable

In the maximization LP problem the change in the coefficient, say c_k, of a basic variable x_k affects the $c_j - z_j$ values corresponding to all non-basic variables in the simplex table. It is because the coefficient c_k is listed in the c_B column of the simplex table and affects the calculation of z_j values.

The sensitivity limits for the contribution per unit of a basic variable are calculated as under:

Lower limit = Original value, c_k – {Lowest absolute value of improvement ratio or $-\infty$ (if no ratio is negative)}

Upper limit = Original value, c_k + {Lowest positive value of improvement ratio or ∞ (if no ratio is positive)}

$$\text{where Improvement ratio} = \frac{\text{Per unit improvement value}}{\text{input - output coefficient in the variable row}}$$

$$= \frac{c_j - z_j}{a_{kj}}$$

Remark: While performing sensitivity analysis, the *artificial variable columns in the simplex table are ignored.* Any exchange of quantities between basic variables and an artificial variable make no sense, because an artificial variable has no economic interpretation. Thus, improvement ratios using coefficients corresponding to artificial variables should not be considered.

Case III : Change in coefficient of a non-basic variable in cost minimization problem

The procedure for calculating sensitivity limits to a cost minimization LP problem, where the objective function coefficients are unit costs is identical to the Case I discussed above. In this case, the unit cost coefficient can be increased to any arbitrary level but it cannot be decreased by more than per unit inprovement value without making it eligible so that a non-basic variable can be entered into the new solution mix. The sensitivity limits can be calculated as:

Lower limit = Original value – Unit improvement value

Upper limit = Infinity (∞)

Alternative Method

If we add Δc_k to the objective function coefficient c_k of the basic variable x_k, then its coefficient c_{Bk} will become $c^*_{Bk} = c_{Bk} + \Delta c_{Bk}$. Now $c_j - z_j$ is calculated as follows:

$$c_j - z_j^* = c_j - \sum_{i=1}^{m} c_{Bi} y_{ij} = c_j - \left\{ \sum_{i \neq k}^{m} c_{Bi} y_{ij} + \left(c^*_{Bk} + \Delta c_{Bk} \right) y_{kj} \right\}$$

$$= c_j - \left\{ \sum_{i \neq k}^{m} c_{Bi} y_{ij} + c_{Bk}\, y_{kj} \right\}$$

$$= c_j - \{z_j + \Delta c_{Bk} y_{kj}\} = (c_j - z_j) - \Delta c_{Bk} y_{kj}$$

For a current basic feasible solution to remain optimal for a maximization LP problem, we must have $c_j - z^*_j \leq 0$. That is,

$$c_j - z_j^* = (c_j - z_j) - \Delta c_{Bk} y_{kj} \leq 0$$

or $$c_j - z_j \, \Delta \leq c_{Bk} y_{kj}$$

or $$\begin{cases} (c_j - z_j)/y_{kj} \leq \Delta c_{Bk} \text{ when } y_{kj} > 0 \\ (c_j - z_j)/y_{kj} \geq \Delta c_{Bk} \text{ when } y_{kj} > 0 \end{cases}$$

Hence, the value of Δc_{Bk} which satisfies the optimality criterion can be determined by solving the following system of linear inequalities:

$$\text{Min} \left\{ \frac{c_j - z_j}{y_{kj} < 0} \right\} \geq \Delta c_{Bk} \geq \text{Max} \left\{ \frac{c_j - z_j}{y_{kj} > 0} \right\} \quad \text{...(1)}$$

Here it may be noted that y_{kj},'s are entries in the non-basic variable columns (i.e. variables) of the optimal simplex table.

The new value of the objective function becomes

$$Z^* = \sum_{i \neq k}^{m} c_{Bi} x_{Bi} + (c_{Bk} + \Delta c_{Bk}) x_{Bk} = \sum_{i \neq k}^{m} c_{Bi} x_{Bi} + \Delta c_{Bk} x_{Bk} = Z + \Delta c_{Bk} x_{Bk}$$

Hence if Δc_{Bk} satisfies inequality (1), then the optimal solution will remain unchanged but the value of Z will be improve by an amount $\Delta c_{Bk} x_{Bk}$..

CHANGE IN THE AVAILABILITY OF RESOURCES (b_1)

Case I : When slack variable is not in the solution mix

We know that in the optimal simplex table, $c_j - z_j$ numbers (ignoring negative sign) corresponding to the slack variable columns represent shadow prices of the available resources. Shadow pricing provides important information in terms of change in the value of objective function from an increase of one unit of a scarce resource.

Another important information which can be obtained from shadow pricing is in terms of resource value that should be increased in order to realize the best marginal increase in the value of the objective function.

The optimal simplex Table 3.1 is reproduced as Table 3.1. Recall that slack variable s_1 represents slack availability of manpower resource and s_2 the Unused raw material. We cannot add an unlimited number of units of resource without violating one of the problem constraints.

For example, once we understand and compute the shadow price for an additional hour of manpower time ($c_4 - z_4$ = Rs. 10/3), we need to determine how many hours we can actually add or remove from manpower so as to increase profit and still have the same shadow price. This process in linear programming involves determining the range over which shadow prices will remain valid.

In Table 1, there is no slack variable in the solution mix column B. The procedure for finding the range for 'resource values' within which current optimal solution remains unchanged is summarized below.

1. Treat the slack variable corresponding to resource value as if it was an entering variable in the solution mix. For this, calculate exchange ratio (minimum ratio) for every row.

$$\text{Exchange ratio} = \frac{\text{Solution value, } x_B}{\begin{array}{c}\text{Exchange(input–Output) coefficients}\\ \text{in a slack variable column}\end{array}}$$

2. Find both lower and upper sensitivity limits.

Lower limit = Original value - Least (smallest) positive ratio or $-\infty$ (if no ratio is positive)

Upper limit = Original value + Smallest absolute negative ratio or ∞ (if no ratio is negative)

Table 1

		$c_j \rightarrow$	4	6	2	0	0
C_B	*Variables in Basis*	*Solution Values*	x_1	x_2	x_3	s_1	s_2
		B	b (= x_B)				
4	x_1	1	1	0	–1	4/3	–1/3
6	x_2	2	0	1	2	–1/3	1/3
Z = 16		z_j	4	6	8	10/3	2/3
		$c_j - z_j$	0	0	–6	–10/3	–2/3

To illustrate the method of finding the range of variation in the availability of resources, we repeat s_1 column and the *Solution values* column from Table 1 and calculate ratios as shown below:

Variables in Basis B	*Solution Values b (= x_B)*	*Exchange Coefficients in s_1-column*	*Exchange Ratio*	*Exchanges Coefficients in s_2-column*	*Exchange Ratio*
x_1	1	4/3	1/(4/3)=3/4	–1/3	1/(–1/3)=–3
x_2	2	–1/3	2/(–1/3)=–6	1/3	2/(1/3)=6

The smallest positive ratio (3/4) indicates as to by how many hours can manpower time resource be decreased (reduced) while smallest absolute negative ratio indicates as to by how many hours can this resource be increased (added) without changing the current optimal product mix. Thus, the shadow price for manpower resource (Rs. 10/3) and raw material resource (Rs. 2/3) are valid over the range as given below:

Solution Mix	Lower Limit	Upper Limit
Manpower (x_1)	3 – (3/4) = 9/4	3 + 6 = 9
Raw material (x_2)	9 – 6 = 3	9 + 6 = 15.

Case II : When a slack variable is in the basis (column B)

When a slack variable is present in solution mix column B of optimal simplex table, the procedure for finding the range of variation for the corresponding right hand of the constraint is as follows:

Lower limit = Original value ù Solution value of slack variable

Upper limit = Infinity (∞)

Case III : Changes in right-hand side when constraints are mixed type

(i) When surplus variable is not in the basis (column B)

Lower limits = Original value – Smallest absolute value of negative exchange ratios or – ∞ (if no ratio is negative)

Upper limit = Original value + Smallest positive minimum ratio or (if no ratio is positive)

(ii) When surplus variable is in the basis (column B)

Lower limit = Minus infinity (– ∞)

Upper limit = Original value + Solution value of surplus variable.

Alternative Methods

1. Since b_1 values are not associated with the calculation of $c_j - z_j$ values, therefore, any change in the right hand-side of the constraints does not affect the optimality condition $[c_j - z_j = c_j - c_B B^{-1} a_j \leq 0$ (maximization case)]. However, it affects the values of basic variables and the value of the objective function $Z (= c_B x_B)$ because in the determination of solution values, $(x_B = B^{-1} b)$, value of resource (i.e. b) is involved. Thus, if resource b_k is changed to $b_k + \Delta b_k$, then the new values of resources becomes

$$\begin{aligned} b^* &= (b_1, b_2, ..., b_k + \Delta b_k, ..., b_m) \\ &= (b_1, b_2, ..., b_m) + (0, 0, ..., \Delta b_k, ..., 0) \\ &= b + \Delta b_k \end{aligned}$$

The range of values within which Δb_k can vary without affecting the optimality of the current solution is determined as follows:

$$\begin{aligned} x^*_B &= B^{-1}(b + \Delta b) = B^{-1} + B^{-1}(\Delta b) \\ &= B^{-1}b + B^{-1}(0, 0, ... \Delta b_k, ..., 0) \\ &= x_B + (\beta_1, \beta_2, ..., \beta_k)(0,0, ... \Delta b_k, ..., 0) \\ &= x_B + \beta_k(\Delta b_k) \end{aligned}$$

$$x^*_{Bi} = x_{Bi} + \beta_{ik}(\Delta b_k);\text{ for ith basic variable}$$

where β_{ik} is the (i, k) element of B^{-1} and is the kth column vector of B^{-1}. In order to maintain the feasibility of the solution at each iteration, the solution values, x_B must be non-negative. That is, we must have

$$x^*_B = x_{Bi} + \beta_{ik}(\Delta b_k) \leq 0;\ i = 1, 2, ..., m$$

Hence, the range of variation in b_k can be obtained by solving the following system of inequalities

$$\min\left\{\frac{x_{Bi}}{\beta_{ik} < 0}\right\} \geq \Delta b_k \text{ Max}\left\{\frac{-x_{Bi}}{\beta_{ik} > 0}\right\}$$

2. The range of variation in the availability of resources (b_i), can also be obtained by using condition of feasibility of the current optimal solution, i.e. $x_B = B^{-1} b \geq 0$,

where B^{-1} = matrix of coefficients corresponding to slack variables in the optimal simplex table

βb_k = amount of change in the resource k

x_B = basic variables appearing in B-column of simplex table.

Remarks:

1. If one or more entries in the x_B-column of the simplex table are negative, the dual simplex method can be used to get an optimal solution to the new problem by maintaining feasibility.
2. A resource whose shadow price is bigger in comparison to others, should be increased first to ensure the best marginal increase in the objective function value.

CHANGES IN THE INPUT-OUT COEFFICIENTS (a_{ij}'s)

Suppose that the elements of coefficient matrix A are changed. Then two cases arise

(i) Change in a coefficient, when variable is a basic variable, and

(ii) Change in a coefficient, when variable is a non-basic variable.

Case I : When a non-basic column $a_k \notin B$ changed to a^*_k, the only effect of such change will be on the optimality condition. Thus the solution will remain optimal, if

$$c_k - z^*_k = c_k - c_B B^{-1} a^*_k \leq 0$$

otherwise the simplex method is continued, after column k of the simplex table is updated, by introducing the non-base variable x_k into the basis. However, the range for the discrete change Δa_{ij} in the coefficient of non-basic variable x_j in the constraint, i can be determined by solving following linear inequalities:

$$\min \left\{ \frac{c_j - z_j}{c_B \beta_i > 0} \right\} \leq \Delta a_{rj} \leq \text{Max} \left\{ \frac{c_j - z_j}{c_B \beta_i < 0} \right\}$$

Here β_i is the ith column B^{-1}.

If $c_B \beta_i = 0$, then Δa_{ij}. is unrestricted in sign.

Alternative Method

The change in the coefficients (a_{ij}'s) values associated with non-basic variables in the optimal simplex table can be analysed by forming a corresponding dual constraint from the original set of constraints:

$$\sum_{i=1}^{m} a_{ij}\, y_i \geq c_j; \text{ for } x_j \text{ non-basic variable.}$$

The values of dual variable y_i's can be obtained from the optimal simplex table. The reason behind this dual constraint formulation is that an activity is considered as fully undertaken provided all the marginal values (shadow prices) of its resources become equal to its per unit contribution to total profit.

Case II : Suppose a basic variable column $a_k \in B$ is changed to a^*_k Then conditions to maintain both feasibility and optimality of the current optimal solution are:

(a) $$\underset{k \neq p}{\text{Max}}\left\{\frac{-x_{Bk}}{x_{Bk}\beta_{pi} - x_{Bp}\,\beta_{ki} > 0}\right\} \leq \Delta a_{ij} \leq \underset{k \neq p}{\text{Min}}\left\{\frac{-x_{Bk}}{x_{Bk}\beta_{pi} - x_{Bp}\,\beta_{ki} < 0}\right\}$$

(b) $$\text{Max}\left\{\frac{c_j - z_j}{(c_j - z_j)\beta_{pi} - y_{pj}\, c_B\, \beta_i > 0}\right\} \leq \Delta a_{ij} \leq \text{Min}\left\{\frac{c_j - z_j}{(c_j - z_j)\beta_{pi} - y_{pj}\, c_B\, \beta_i < 0}\right\}$$

SENSITIVITY ANALYSIS IN ASSIGNMENT PROBLEMS

There is very little scope for sensitivity analysis in assignment problem because of its structure. Modest alterations in the conditions (such as one being able to do two jobs) can be considered by repeating the man's row and adding a dummy a column to square up the matrix. The problem of not assigning a particular job (ith) to particular facility (jth) can be solved by taking a very large cost (∞) of this assignment.

Also, the addition of a constant throughout any row or column does not affect the optimal solution of the assignment problem.

THE TRAVELLING SALESMAN (ROUTING) PROBLEM

Suppose a salesman wants to visit a certain number of cities.

He knows the distances (or time or cost) of journey between every pair of cities allotted to him. His problem is to choose such a rout which starts from his home city, passes through each city once and only once and returns this home city in the shortest possible distance (or in least time or at least cost).

The above problem may be classified in two forms:

(i) Symmetrical

If the distance (or time or cost) between every pair of cities is independent of the direction of journey the problem is said to be symmetrical.

(ii) Asymmetrical

If for one or more pairs of cities, the distance (or time or cost) changes with the direction, the problem is said to be asymmetrical. For example, it takes longer time while going up-hill from city A to B instead of coming down hill from city B to A. Similarly flying from East to West usually takes longer time than from West to East on account of prevailing winds.

Further we note that for two cities there is only one possible route *i.e.*, there is no choice.

In case of three cities, say A, B and C, one of them (say A) is the home base, there are two possible routes: A → B → C and A → C → B. For four cities there are 3! = 6 possible routes. In general, there are (n – 1)! possible routes if there are n cities. Thus, practically it is impossible to find the best route by trying each one. That is why the travelling salesman problem is considered as a puzzle by the mathematicians. The best procedure to solve the problem is as if it were an assignment problem. We formulate the problem of travelling salesman in the form of an assignment problem with the additional restriction on his choice of route.

FORMULATION OF A TRAVELLING SALESMAN PROBLEM AS ASSIGNMENT PROBLEM

Let x_{ij} = 1, if the salesman goes directly from city A_j, to c_{ij} to A_j and zero otherwise. Also, let c_{ij} be the distance (or time or cost) from city A_i to city A_j. Then our problem is to minimize $Z = \sum\sum c_{ij}\, x_{ij}$ with one additional restriction that the x_{ij}'s must be so chosen that no city is visited i_j twice until the tour of all the cities is completed. In particular, he cannot go directly from city A_i to A_i itself. To avoid this possibility in the minimization process we adopt the convention $c_{ij} = \infty$ so that x_{ij} can never be unity when i = j. Also we note that only one x_{ij} = 1 for each value of i and j. The distance (or time or cost) matrix for this problem is given in the following table:

To

From		A_1	A_2	...	A_n
	A_1	∞	c_{12}	...	c_{1n}
	A_2	c_{21}	∞	...	c_{2n}
	.	.	.	.	.
	.	.	.	.	.
	.	.	.	.	.
	A_n	c_{n1}	c_{n2}	...	∞

We can omit the variable x_{ij} from the problem specification. Our problem is to determine a set of n elements of this matrix, one in each row and one in each column, so as to minimize the sum of the elements determined above.

Note: A problem similar to travelling salesman arises when n items say A_i, i = 1, 2, ... n, are to be produced on a machine in continuation, given that c_{ij} (i, j = 1, 2, ... n) is the setup cost of the machine when item A_i is followed by A_j. Here two additional restrictions are imposed. One restriction is that we do not follow A_i again by A_i. The other restriction is that we do not produce an item again until all items are produced once.

Solution Procedure

The problem could be solved by assignment technique. In some cases (violation of additional restriction) we use the method of numeration by assignment the next minimum element of matrix in place of zero.

ADDITION OF A NEW CONSTRAINT TO THE PROBLEM

Let a new constraint be introduced to a L.P.P. whose optimal solution has been obtained. Here we assume that the additional constraint does not introduce any new variable with non-zero-price.

Let Z* be the optimal (maximal) value of the objective function for the new L.P.P. while it was Z for the original problem. Let Z* > Z. Since the new optimal solution satisfies the first m constraints as well as the additional new constraint, therefore, it is also an optimal solution of the original problem which is contradiction to the fact that we already had an optimal solution to the original L.P.P. Here $Z^* \leq Z$. (Max.)

Thus, we have the following two cases:

Case I : If the optimal solution of the original L.P.P. satisfies the new constraint, it is also an optimal solution of the new L.P.P. In this case the adulterant is redundant.

Case II : If the optimal solution of the original L.P.P. does not satisfy the new constraint, a new optimal solution of the new L.P. problem must be obtained as follows:

To Find the New Optimal Solution of the New Enlarged Problem

Let B and B_1 be the optimal basis of the original and the enlarged L.P. problem respectively. Clearly B_1 is the square matrix of order (m + 1) if B the square matrix of order m.

∴ We can write

$$B_1 = \begin{bmatrix} B & 0 \\ \alpha & \pm 1 \end{bmatrix} \qquad \text{...(1)}$$

The last column of B_1 corresponds to the slack, surplus or artificial vector associated with the additional new constraints and α is the row vector of the coefficients, in the new constraint, of the variables which correspond to the vectors in the optimal basis B.

Since B^{-1} exist and is known therefore the inverse of B_1 is given by

$$B_1^{-1} = \begin{bmatrix} B^{-1} & 0 \\ \mp\alpha B^{-1} & \pm 1 \end{bmatrix} \qquad \text{...(2)}$$

Let a_{m+1}, j be the coefficient of x_j in the new (m + 1)th constraint and α_j^* the column vector of the coefficients of x_j in the enlarged problem. Also if Y_j^* and Z_j^* are Y_j and Z_j for the new problem, then we have

$$Y_j^* = B_1^{-1}.\alpha_j^* = \begin{bmatrix} B^{-1} & 0 \\ \mp\alpha B^{-1} & \pm 1 \end{bmatrix} \begin{bmatrix} \alpha_j \\ a_{m+1.j} \end{bmatrix}$$

$$= \begin{bmatrix} Y_j \\ \mp\alpha B^{-1}\alpha_j \pm a_{m+1,j} \end{bmatrix}$$

$$\Rightarrow \quad Y_j^* = \begin{bmatrix} Y_j \\ \mp\alpha Y_j \pm a_{m+1,j} \end{bmatrix}$$

Now $Z_j^* = c_{B1}\ Y_j^* = (c_B, c_{B.m+1}) \begin{bmatrix} Y_j \\ \mp\alpha Y_j \pm a_{m+1,j} \end{bmatrix}$

or $\quad Z_j^* = c_{B1}\ Y_j + c_{B,\ m+1}.\ (\pm\ \alpha Y_j \pm a_{m+1,j})$ ...(3)

(i) *If slack or surplus variable is introduced in the additional constraint.*

In this case $c_{B,m+1} = 0$

$\therefore$ from (3), $Z_j^* = c_B\ Y_j = Z_j$

$\therefore \quad c_j - Z_j^* = c_j - Z_j$

$\therefore \quad c_j - Z_j^*$

remains unchanged is this case. Since the optimal solution of the original problem does not satisfy the new constraint, therefore the slack or surplus variable introduced in the new constraint is negative. *Hence we can apply the dual simplex algorithm to find an optimal feasible solution of the new problem.*

(ii) *If the artificial variable is introduced in the additional constraint i.e., if the additional constraint is a perfect equality.* In this case the additional vector is an artificial vector. Now there are two possibilities:

(a) *If the artificial variable in the basic solution is negative,* then assigning a price to the artificial variable we can use the dual simplex algorithm for the removal of the artificial variable from the basis.

(b) *if the artificial variable in the basis solution is positive,* then assigning a price –M to the artificial variable we can use the standard simplex method for the removal of the artificial variable from the basis. It is important to note that in this case $c_j - Z_j$ will be changed.

VARIATION IN THE COMPONENT a_{ij} OF THE COEFFICIENT MATRIX A

Let the element a_{lk} of the *l*-th row and k-th column of the coefficient matrix A be changed to $a_{lk} + \Delta a_{lk}$. Now there are two possibilities according as a_{lk} is or not the element of the optimal basis B.

Case I. When a_{lk} is not an Element of Optimal Basis B

If a_{lk} is not an element of the optimal basis B, a change in a_{lk} will not affect B and so B^{-1} remains the same. Thus the change in such a_{lk} will not affect the solution $x_B = B^{-1}b$. Hence the feasibility of the solution x_B is not affected. But the change of a_{lk} may change the optimality of the solution *i.e.*, the solution x_B may not be the optimal solution of the new L.P.P. (obtained by changing a_{lk} to $a_{lk} + \Delta a_{lk}$). Thus we have to find the range of variation of a_{lk} so that the solution still remains optimal.

Since all the column voters except a_k of the matrix A remain unaffected by the variation Δa_{lk} in a_{lk}.

$\therefore\ \Delta_j = c_j - Z_j \le 0$ for all j ($\neq$ k) not in the basis.

Thus the solution x_B will remain optimal for the new L.P.P. also, if

$$\Delta_k^* = c_k - Z_k^* \le 0 \qquad ...(1)$$

where Δ_k^* and Z_k^* are Δ_k and Δ_k for the new L.P.P.

Let $B^{-1} = (\beta_1, \beta_2, \ldots, \beta_l, \ldots, \beta_m)$

and $\alpha_k = [a_{lk}, a_{2k}, \ldots, a_{lk}, \ldots, a_{mk}]$

If α_k^* is α_k for the new L.P.P. then

$$\alpha_k^* = [a_{lk}, a_{2k}, \ldots, a_{lk} + \Delta a_{lk}, \ldots, a_{mk}]$$
$$= [a_{1k} + 0, a_{2k} + 0, \ldots, a_{lk} + \Delta_{lk}, \ldots, a_{mk} + 0]$$
$$= [a_{1k}, a_{2k}, \ldots, a_{lk}, \ldots, a_{mk}] + [0, 0, \ldots, \Delta a_{lk}, \ldots, 0]$$
$$= \alpha_k + [0, 0, \ldots, \Delta a_{lk}, \ldots, 0]$$

l-th component

and $Z_k{}^* = c_B\ B^{-1}\ \alpha_k{}^*$

$= c_B .\ B^{-1}\ \{\alpha_k + [0, 0, \ldots, \Delta a^{lk}, \ldots 0]\}$

$= c_B .B^{-1}\ \alpha_k + c_B .B^{-1}\ [0, 0, \ldots, \Delta a_{lk}, \ldots, 0]$

$= Z_k + c_B .(\beta_1, \beta_2, \ldots, \beta_l, \ldots, \beta_m).[0, 0, \ldots \Delta a_{lk}, \ldots, 0]$

$= Z_k + c_B\ (\beta_l\ \Delta a_{lk})$

$= Z_k + \Delta a_{lk} .\ c_B\ \beta_l$

$= Z_k + \Delta a_{lk}\ (c_{B1}, c_{B2}, \ldots, c_{Bl}, \ldots, c_{Bm}) \times [\beta_{1l}, \beta_{2l}, \ldots, \beta^{ll}, \ldots \beta_{ml}]$

$= Z_k + \Delta a^{lk} \sum_{i=1}^{m} c_{Bi} \cdot \beta_{il}\ .$

$\therefore$ From (1), the solution x_B will remain optimal if

$$\Delta^{k*} = c_k - Z_k{}^* = c_k - Z_k - \Delta a_{lk} \sum_{i=1}^{m} c_{Bi} \cdot \beta_{il} \le 0$$

or $\quad \Delta a^{lk} \sum_{i=1}^{m} c_{Bi} \cdot \beta_{il}\ {}^3\ c^k - Z^k\ (= \Delta_k)$

$$\therefore \quad \Delta a_{lk} \ge \frac{\Delta_k}{\sum_{i=1}^{m} c_{Bi} \beta_{il}} \quad \text{for } \sum_{i=1}^{m} c_{Bi} \beta_{il} > 0.$$

$$\text{and } \Delta a_{lk} \le \frac{\Delta_k}{\sum_{i=1}^{m} c_{Bi} \beta_{il}} \quad \text{for } \sum_{i=1}^{m} c_{Bi} \beta_{il} < 0.$$

Hence the range of Δa_{lk} (change in $a_{lk} \notin B$), so that the solution x_B remains optimal and feasible, is given by

$$\left[\frac{\Delta_k}{\left(\sum_{i=1}^{m} c_{Bi} \beta_{il}\right) > 0}\right] \le \Delta a_{lk} \le \left[\frac{\Delta_k}{\left(\sum_{i=1}^{m} c_{Bi} \beta_{il}\right) < 0}\right] \qquad \ldots(2)$$

If $\sum_{i=1}^{m} c_{Bi} \cdot \beta_{il} = 0$, $\Delta\ a_{lk}$ is unrestricted.

If $\sum_{i=1}^{m} c_{Bi} \cdot \beta_{il} > 0$, there is no upper bound to $\Delta\ a_{lk}$,

and if $\sum_{i=1}^{m} c_{Bi} \cdot \beta_{il} < 0$, there is no lower bound to $\Delta\ a_{lk}$.

Note: There is no change in the value of the objective function if D a_{lk} satisfies (2).

Case II. When a_{lk} is an Element of the Optimal Basis B

Since a_{lk} is an element of the optimal basis B then if a_{lk} is changed to $a_{lk} + \Delta a_{lk}$, the optimal basis B will certainly be changed and hence B^{-1} will also be changed. Consequently $x_B = B–1$ b and $Zj = c_B B^{-1}\alpha_j$ will also change. A change in Z_j may disturb the optimality condition $\Delta_j = c_j - Z_j \le 0$ while a change in x_B may disturb the feasibility of the solution. *Hence our aim is to find the range of variation of a_{lk} so that neither the optimality nor the feasibility of the solution is disturbed.*

When the element $a_{lk} \in B$ is changed to $a_{lk} + \Delta a_{lk}$ then let the optimal basis B, the solution x_B and Z_j be represented by B*, x_B* and Z^*_j respectively.

Let $B = (b_1, b_2, \ldots, b_m)$ and $\alpha_k = bp$

$\therefore \quad a_{lk} = b_{lp}$ and $a_{lk} + \Delta a_{lk} = b_{lp} + \Delta b_{lp}$

$$\therefore \quad B^* = B = \begin{pmatrix} 0 & 0 & \cdots & 0 & \cdots & 0 \\ 0 & 0 & \cdots & 0 & \cdots & 0 \\ \vdots & \vdots & & \vdots & & \vdots \\ 0 & 0 & \cdots & \Delta b_{lp} & \cdots & 0 \\ \vdots & \vdots & & \vdots & & \vdots \\ 0 & \cdots & 0 & \cdots & 0 & 0 \end{pmatrix} \leftarrow l-\text{th row}.$$

p-th column

$= B + \Delta b_{lp}.\ 0_{lp}$, where 0_{lp} is the null matrix except for the (l, p)th element which equals to unity

$= B\ (I + B^{-1}\ \Delta\ b_{lp}.\ 0_{lp})$

$$\therefore \quad B^{*-1} = \{B(I + B^{-1}\ \Delta b_{lp}.\ 0_{lp})\}-1 = (I + B^{-1}\ \Delta b_{lp}.\ 0_{lp})^{-1}.\ B^{-1} \qquad \ldots(3)$$

$[\because (AB)^{-1} = B^{-1} A^{-1}]$

If $B^{-1} = (\beta_1, \beta_2, \ldots, \beta_m)$,

$$I + B^{-1}\ \Delta b_{lp}\ .0_{lp} = I + B^{-1} \begin{pmatrix} 0 & 0 & \cdots & 0 & \cdots & 0 \\ 0 & 0 & \cdots & 0 & \cdots & 0 \\ \vdots & \vdots & & \vdots & & \vdots \\ 0 & 0 & \cdots & \Delta b_{lp} & \cdots & 0 \\ \vdots & \vdots & & \vdots & & \vdots \\ 0 & \cdots & 0 & \cdots & 0 & 0 \end{pmatrix}$$

$$= I + \xrightarrow[\text{p-th row}]{} \begin{bmatrix} \beta_{11} & \beta_{12} & \cdots & \beta_{1l} & \cdots & \beta_{lm} \\ \beta_{21} & \beta_{22} & \cdots & \beta_{2l} & \cdots & \beta_{2m} \\ \vdots & \vdots & & \vdots & & \vdots \\ \beta_{pl} & \beta_{pl} & \cdots & \beta_{pl} & \cdots & \beta pm \\ \vdots & \vdots & & \vdots & & \vdots \\ \beta_{ml} & \beta_{m2} & \cdots & \beta_{ml} & \cdots & \beta_{mm} \end{bmatrix}$$

$$\times \begin{pmatrix} 0 & 0 & \cdots & 0 & \cdots & 0 \\ 0 & 0 & \cdots & 0 & \cdots & 0 \\ \vdots & \vdots & & \vdots & & \vdots \\ 0 & 0 & \cdots & \Delta b_{lp} & \cdots & 0 \\ \vdots & \vdots & & \vdots & & \vdots \\ 0 & \cdots & 0 & \cdots & 0 & 0 \end{pmatrix} \begin{matrix} \\ \\ \\ \leftarrow \text{l-th row} \\ \\ \\ \end{matrix}$$

(l-th column)

$$= \begin{matrix} \\ \\ \\ \text{p-th row} \rightarrow \\ \\ \\ \end{matrix} \begin{pmatrix} 1 & 0 & \cdots & 0 & \cdots & 0 \\ 0 & 1 & \cdots & 0 & \cdots & 0 \\ \vdots & \vdots & & \vdots & & \vdots \\ 0 & 0 & \cdots & 1 & \cdots & 0 \\ \vdots & \vdots & & \vdots & & \vdots \\ 0 & 0 & \cdots & 0 & \cdots & 0 \end{pmatrix}$$

(p-th column)

$$+ \begin{pmatrix} 0 & 0 & \cdots & \beta_{1l}\Delta b_{lp} & \cdots & 0 \\ 0 & 0 & \cdots & \beta_{2l}\Delta b_{lp} & \cdots & 0 \\ \vdots & \vdots & & \vdots & & \vdots \\ 0 & 0 & \cdots & \beta_{pl}\Delta b_{lp} & \cdots & 0 \\ \vdots & \vdots & & \vdots & & \vdots \\ 0 & 0 & \cdots & \beta_{ml}\Delta b_{lp} & \cdots & 0 \end{pmatrix} \begin{matrix} \\ \\ \\ \leftarrow \text{p-th row} \\ \\ \\ \end{matrix}$$

(p-th column)

$$\Rightarrow \quad x_B{}^* = \begin{bmatrix} x_{B1}\,0\dfrac{\beta_{1l}\,.\,\Delta b_{lp}}{D}\,.\,x_{Bp} \\ x_{B2} - \dfrac{\beta_{21}\,.\,\Delta b_{lp}}{D}\,.\,x_{Bp} \\ \vdots \\ \dfrac{1}{D}\,x_{Bp} \\ x_{Bm} - \dfrac{\beta_{ml}\,.\,\Delta b_{lp}}{D}\,.\,x_{Bp} \end{bmatrix} \qquad \ldots(4)$$

(p-th column)

The solution $\mathbf{x}_B{}^*$ is feasible if $\mathbf{x}_B{}^* \geq 0$.

$\therefore$ From (4), the solution $x_B{}^*$ is feasible,

$$\text{if} \quad x_{Bi} - \frac{\beta_{il}\,.\,\Delta b_{lp}}{D}\,.\,x_{Bp} \leq 0 \qquad \ldots(5)$$

for all i = 1, 2, ... ,m (i ≠ p)

and $\frac{1}{D} . x_{Bp} \geq 0.$ ...(6)

Since $x_{Bp} \geq 0$

$\therefore$ (6) will hold of we have

$\therefore$ from (5), (assuming that (6) is true), we have

$x_{Bi}\ D - \beta_{il} . \Delta b_{lp}\ x_{Bp} \geq 0$

$\Rightarrow x_{Bi} . (1 + \beta_{pl}\ \Delta b_{lp}) - \beta_{il}\ \Delta b_{pl}\ x_{Bp} \geq 0$

$\Rightarrow x_{Bi} \geq (\beta_{il}\ x_{Bp} - \beta_{pl}\ x_{Bi}) . \Delta b_{lp}$ for all $i \neq p$

$\therefore \Delta\ b_{lp} \leq \frac{x_{Bi}}{\beta_{il} x_{Bp} - \beta_{pl} x_{Bi}}$, for $\beta_{il}\ x_{Bp} - \beta_{pl}\ x_{Bi} < 0$

and $\Delta\ b_{lp} \geq \frac{x_{Bi}}{\beta_{il} x_{Bp} - \beta_{pl} x_{Bi}}$, for $\beta_{il}\ x_{Bp} - \beta_{pl}\ x_{Bi} < 0.$

Hence the range of $\Delta\ a_{lk}\ (= \Delta\ b_{lp}) \in B$, so that the solution remains feasible, is given by

$$\text{Max.}\left[\frac{x_{Bi}}{P_i < 0}\right] \leq \Delta a_{lk} \leq \text{Min}\left[\frac{x_{Bi}}{P_i > 0}\right] \quad ...(7)$$

where $D = 1 + \beta_{pl}\ \Delta\ \beta_{lp}, = 1 + \beta_{pl}\ \Delta\ a_{lk} > 0.$

$Pi = \beta_{il}\ x_{Bp} - \beta_{pl}\ x_{Bi}.$

If no $P_i < 0$, there is no lower bound to $\Delta\ a_{lk}$ and if no $P_i > 0$, there is no upper bound to $\Delta\ a_{lk}$.

To Find the Range of Variation of a_{lk} for the Optimality of the Solution

For optimality of the solution x_B*, we must have

$\Delta_j^* = c_j - Z_j^* \leq 0$

For all j not in the basis.

$Z_j^* = c_B\ B^{*-1}\ \alpha_j = c_B\ (I + B^{-1} .\ \Delta\ b_{lp}\ 0_{lp})^{-1} .\ B^{-1}\ \alpha_j$

$= c_B\ (I + B^{-1} .\ \Delta\ b_{lp}\ 0_{lp})^{-1} .\ Y_j$

Since $B^{-1}\ \alpha_j = Y_j = [y_{1j}, y_{2j}, ... ,y_{pj}, ... ,y_{mj}].$

Using (4) and simplifying as before as in (1), we have

$$Z_j^* = c_B \begin{bmatrix} y_{1j} - \dfrac{\beta_{1l}\Delta b_{lp}}{D}.y_{pj} \\ y_{2j} - \dfrac{\beta_{2l}\Delta b_{lp}}{D}.y_{pj} \\ \vdots \\ \dfrac{1}{D}y_{pj} \\ \vdots \\ y_{mj} - \dfrac{\beta_{ml}\Delta b_{lp}}{D}.y_{pj} \end{bmatrix}$$

$$= (c_{B1}, c_{B2}, \ldots, c_{Bp}, \ldots, c_{Bm}). \begin{bmatrix} y_{1j} - \dfrac{\beta_{1l}\Delta b_{lp}}{D}.y_{pj} \\ y_{2j} - \dfrac{\beta_{2l}\Delta b_{pj}}{D} \\ \vdots \\ \dfrac{1}{D}y_{pj} \\ y_{mj} - \dfrac{\beta_{ml}\Delta b_{lp}}{D}.y_{pj} \end{bmatrix} \xleftarrow[\text{p-th row}]{lp.y_{pj}}$$

$$= \sum_{i=1}^{m} c_{Bi}\left(y_{ij} - \frac{\beta_{il}\Delta b_{lp}}{D}y_{pj}\right) + \frac{c_{By}.y_{pj}}{D}$$

$$= \sum_{i=1}^{m}\left(c_{Bi}y_{ij} - \frac{c_{Bi}\beta_{il}\Delta b_{lp}}{D}y_{pj}\right) - c_{Bp}\left(y_{pj} - \frac{\beta_{pl}\Delta b_{lp}}{D}y_{pj}\right) + \frac{c_{Bp}.y_{pj}}{D}$$

$$= \sum_{i=1}^{m} c_{Bi}y_{ij} - \frac{1}{D}\sum_{i=1}^{m} c_{Bi}\beta_{il}\Delta b_{lp}y_{pj} - \frac{1}{D}[c_{Bp}.(D - b_{pl}\,\Delta\, b_{lp})\,y_{pj} - c_{Bp}\,y_{pj}]$$

$$= Z_j - \frac{1}{D}\sum_{i=1}^{m} c_{Bi}\beta_{il}\Delta b_{lp}y_{pj}$$

Since $\sum_{i=1}^{m} c_{Bi}y_{ij} = Z_j$ and $D = 1 + \beta_{pl}\,\Delta b_{lp}$

$\therefore$ For optimality of the solution x_B^*

$$\Delta_j^* = c_j - Z_j^* = c_j - Z_j + \frac{1}{D}\sum_{i=1}^{m} c_{Bi}.\beta_{il}\Delta b_{lp}y_{pj} \le 0 \;\forall\; j \text{ not in the basis}$$

$$\Rightarrow D\,(c_j - Z_j) + \sum_{i=1}^{m} c_{Bi}\beta_{il}\Delta b_{lp}y_{pj} \le 0 \text{ assuming that } D = 1 + b_{pl}\,\Delta\, b_{lp} > 0$$

$$\Rightarrow (1 + b_{pl}\ \Delta b_{lp})\ (c_j - Z_j) + \sum_{i=1}^{m} c_{Bi}\beta_{il}\Delta b_{lp} y_{pj} \le 0$$

$$\Rightarrow \left[\beta_{pl}(c_j - Z_j) + \sum_{i=1}^{m} c_{Bi}\beta_{il} y_{pj}\right] \Delta\ b_{lp} \le -\ (c_j - Z_j)$$

$$\Rightarrow Q_j\ \Delta\ b_{lp} \le -\ \Delta_j$$

where $Q_j = b_{pl}\ (c_j - Z_j) + \sum_{i=1}^{m} c_{Bi}\beta_{il} y_{pj}$

$$= b_{pl}\ \Delta_j + \left[\sum_{i=1}^{m} c_{Bi}\beta_{il}\right] y_{pl}$$

$$\therefore \qquad \Delta\ b_{lp} \le \frac{-\Delta_j}{Q_j},\ \text{for } Q_j > 0$$

and $\Delta\ b_{lp} \ge \dfrac{-\Delta_j}{Q_j}$, for $Q_j < 0$.

Hence the range of $\Delta\ a_{lk}$ $(= \Delta\ b_{lp})$ *so that the solution remains optimal is given by*

$$\text{Max.} \left[-\frac{\Delta_j}{Q_j < 0}\right] \le \Delta\ a_{lk} \le \text{Min} \left[\frac{-\Delta_j}{Q_j > 0}\right] \qquad ...(8)$$

for all j not in the basis.

If no $Q_j < 0$, there is no lower bound to $\Delta\ a_{lk}$, and if no $Q_j > 0$, there is no upper bound to $\Delta\ a_{lk}$.

Hence a change Δa_{lk} in a_{lk} (an element of basis matrix B) for the solution to remain feasible and optimal can be made such that (6), (7) and (8) are satisfied.

To Find Limits of Variation of a_{1k} for the Feasibility of the Solution

From (9) of 4, the range of variation of $a_{lk} = a_{11}$ for the feasibility of the solution is given by

$$\text{Max.} \left[\frac{x_{Bi}}{P_i < 0}\right] \le \Delta\ a_{lk} \le \text{Min.} \left[\frac{x_{Bi}}{P_i > 0}\right] \qquad ...(1)$$

where $D = 1 + \beta_{pl}\ \Delta\ \beta_{lp} \ge 0$, $P_i = \beta_{il}\ x_{Bp} - \beta_{pl}\ x_{Bi}$.

Here $a_{lk} = a_{11} \in B$. Since $a_{11} \in \alpha_1\ (= Y_1) = y_2$

$\therefore\ \ a_{lk} = a_{11} = b_{12} = b_{lp}$ *i.e.*, l = 1 and k = 1, p = 2

and i = 1, 2, 3. (Since there are only three rows in A)

$x_{B1} = 56/27$, $x_{B2} = 5/3$, $x_{B3} = 5/27$.

$\therefore$ $P_1 = \beta_{11}\, x_{B2} - \beta_{21}\, x_{B1} = (5/9).\ (5/3) - 0 = 25/27 > 0$

$P_2 = \beta_{21}\, x_{B2} - \beta_{21}\, x_{B2} = 0.5/3 - 0 = 0$

$P_3 = \beta_{31}\, x_{B2} - \beta_{21}\, x_{B3} = -\ 1/9.\ 5/3 - 0 = -\ 5/27 < 0.$

$\therefore$ from (1), we have

$$\text{Max.}\left[\frac{x_{B3}}{P_3 < 0}\right] \le \Delta\, a_{11} \le \text{Min.}\left[\frac{x_{B1}}{P_1 > 0}\right]$$

$$\Rightarrow \frac{5/27}{5/27} \le \Delta a_{11} \le \frac{56/27}{25/27}$$

$\Rightarrow -1 \le \Delta\, a_{11} \le 56/25$ ...(A)

Also $D = 1 + \beta_{pl}\, \Delta\, a_{lk} = 1 \ge 0$, holds as $\beta_{pl} = \beta_{21} = 0$.

Again from (10) 4, the range of variation of $a_{lk} = a_{11}$ for the optimality of the solution is given by

$$\text{Max.}\left[\frac{-\Delta_j}{Q_j < 0}\right] \le \Delta a_{lk} \le \text{Min.}\left[\frac{-\Delta_j}{Q_j > 0}\right] \quad \text{...(2)}$$

where $Q_j = \beta_{pl}\, \Delta_j + \left(\sum_{i=1}^{m} c_{Bi}\beta_{il}\right) y_{pj}$ for all j not in the basis

Here $Q_j = \beta_{21}\, \Delta_j + \left(\sum_{i=1}^{3} c_{Bi}\beta_{il}\right) y_{2j}$

$= \beta_{21}\, \Delta_j + (c_{B1}\, \beta_{11} + c_{B2}\, \beta_{21} + c_{B3}\, \beta_{31})\, y_{2j}$

$= 0 + (3.5/9 + 10.0 - 5.1/9)\, y_{2j} = (10/9)\, y_{2j}$ for all j = 3, 5, 6, 7 not in the basis.

$\therefore$ $Q_3 = 10/9.y_{23} = 10/9.2/3 = 20/27 > 0$

$Q_5 = 10/9.y_{25} = 10/9.0 = 0$

$Q_6 = 10/9.y_{26} = 10/9.1/3 = 10/27 > 0$

$Q_7 = 10/9.y_{27} = 10/9.0 = 0.$

Since no $Q_j < 0$

$\therefore$ There is no lower bound to $a_{lk} = (a_{11})$

$\therefore$ from (2), we have

$$-\infty < \Delta\, a_{11} \le \text{Min.}\left\{-\frac{\Delta_3}{Q_3}, -\frac{\Delta_6}{Q_6}\right\}$$

$$\Rightarrow -\infty < \Delta\, a_{11} \le \text{Min.}\left\{\frac{82/27}{20/27}, \frac{80/27}{10/27}\right\}$$

$\Rightarrow -\infty < \Delta\, a_{11} \le 41/10.$ ...(B)

Since the range for Δa_{11} so that the solution remains optimal and feasible is given (A) and (B) both

$\therefore$ Both (A) and (B) are satisfied

if $-1 \le \Delta a_{11} \le 56/25$.

Since $a_{11} = 1$

$\therefore$ limits of variation of a_{11} are

$-1 + 1 \le a_{11} \le 56/25 + 1$

$\Rightarrow 0 \le a_{11} \le 81/25$.

To find limits of variation of a_{23}. Here $a_{23} \notin \alpha_3$ (Y_3) which is not in B. From (2), 4 the range of Δa_{lk} (change in $a_{lk} \notin$ B) so that the solution remains optimal and feasible is given by

$$\left[\frac{\Delta_k}{\left(\sum_{i=1}^{m} c_{Bi}\beta_{il}\right) > 0}\right] \le \Delta a_{lk} \le \left[\frac{\Delta_k}{\left(\sum_{i=1}^{m} c_{Bi}\beta_{il}\right)} < 0\right] \quad ...(3)$$

Here $a_{lk} = a_{23} = 2$

$\therefore$ $l = 2$, $k = 3$.

$\Delta_k = \Delta_3 = -82/27$, $m = 3$ (number of constraints)

$$\sum_{i=1}^{m} c_{Bi}\beta_{il} = \sum_{i=1}^{3} c_{Bi}\beta_{i2} = c_{B1}\,\beta_{12} + c_{B2}\,\beta_{22} + c_{B3}\,\beta_{32}$$

$= 3.\ (-5/27) + 10.\ (1/3) + 5\ (1/27) = 80/27 > 0$

$\therefore$ There is no upper bound to Δa_{23}.

$\therefore$ from (3), we have

$$\frac{\Delta_3}{\sum_{i=1}^{3} c_{Bi}\beta_{i2} > 0} \le \Delta a_{23} < \infty$$

$$\Rightarrow \frac{-82/27}{80/27} \le \Delta a_{23} < \infty$$

$\Rightarrow -41/40 \le \Delta a_{23} < \infty$.

Since $a_{23} = 2$,

$\therefore$ limits of variation of a_{23} are

$-41/40 + 2 \le a_{23} < \infty + 2$

$\Rightarrow 39/40 \le a_{23} < \infty$.

VARIATION IN THE PRICE VECTOR c

Consider the L.P.P Max. $Z = cx$, s.t. $Ax = b$, $x \geq 0$. If x_B is the optimal basic feasible solution and B the optimal basis matrix, we have

$$x_B = B^{-1} b.$$

It is clear that x_B is independent of therefore the change in some component c_j of c will not change x_B, i.e., x_B will always remain basic feasible solution.

Condition of optimality for the solution x_B, is $\Delta_j = c_j - Z_j \leq 0$ for all j not in the basis which is satisfied before any change in c_j. But when c_j changes, the condition of optometry may not be satisfied. The change in the price vector c maybe made in the following two ways:

(i) Variation in $c_j \notin c_B$

(i.e., Change in c_j Which is the Price of the Non-Basic Variable x_j)

Since x_B is an optimal solution of the L.P.P. (maximization problem)

$\therefore \quad \Delta_j = c_j - Z_j \leq 0$, for all j not in the basis.

If Δc_k is the change in the cost c_k and c_k is not present in c_B (the vector of the costs associated to basic variables) then c_B remains unaltered with this change. Also there is no change in B.

$\therefore \quad Z_j = c_B B^{-1} \alpha_j = c_B Y_j$ also remains unaltered.

If x_B is still an optimal solution of the given L.P.P. when c_k changes to $c_k + \Delta c_k$ then, $c_j - Z_j$ remains same for all j ($\neq$ k) not in the basis,

$\therefore$ We must have

$$(c_k + \Delta c_k) - Z_k \leq 0$$

or $$\Delta c_k \leq Z_k - c_k \; (= -\Delta_k). \qquad ...(1)$$

Hence if $c_k \notin c_B$ changes to $c_k + \Delta c_k$ such that $\Delta c_k \leq Z_k - c_k$ $(= -\Delta_k)$, the value of the objective function and the optimal solution of the problem remains unchanged. Note that there is no lower bound to Δc_k.

(ii) Variation in $c_j \in c_B$.

(i.e., Change in c_j Which is the Price of the Basic Variable)

We know that

$$Z_j = c_B B^{-1} a_j = c_B Y_j = \sum_{i=1}^{m} c_{Bi} Y_{ij}.$$

Let Δc_{Bk} be the change in c_{Bk}, the price of the basic variable x_{Bk} then if Z_j^* is the value of Z_j in this solution, then we have

$$Z_j^* = \sum_{\substack{i=1 \\ i \neq k}}^{m} c_{Bi} y_{ij} + (c_{Bk} + \Delta c_{Bk}) y_{kj}$$

$$= \sum_{i=1}^{m} c_{Bi} y_{ij} + y_{kj} \Delta c_{Bk} = Z_j + y_{kj} \Delta c_{Bk}$$

$$\therefore \quad c_j - Z_j^* = c_j - (Z_j + y_{kj} \Delta c_{Bk}) = (c_j - Z_j) - y_{kj} \Delta c_{Bk}.$$

The solution $\mathbf{x}_B$ will remain optimal for the change Δc_{Bk} in c_{Bk} if

$$c_j - Z_j^* \leq 0 \text{ for all j not in the basis}$$

or $(c_j - Z_j) - y_{kj} \Delta c_{Bk} \leq 0$

or $y_{kj}, \Delta c_{Bk} \geq c_j - Z_j$

$$\therefore \quad \Delta c_{Bk} \geq \frac{c_j - Z_j}{y_{kj}}, \text{ for } y_{kj} > 0$$

$$\text{and } \Delta c_{Bk} \leq \frac{c_j - Z_j}{y_{kj}}, \text{ for } y_{kj} < 0. \quad \text{...(2)}$$

Here the range of Δc_{Bk} (change in the price of c_{Bk} corresponding tooth basis variable x_{Bk}) so that the solution remains optimal is given by

$$\underset{y_{kj} > 0}{\text{Max}} \cdot \left[\frac{c_j - Z_j}{y_{kj}}\right] \leq \Delta c_{Bk} \leq \underset{y_{kj} < 0}{\text{Min}} \cdot \left[\frac{c_j - Z_j}{y_{kj}}\right] \quad \text{...(3)}$$

for all j corresponding to which α_j is not in the optimal basis.

If no $y_{kj} > 0$, there is no lower bound to Δc_{Bk} and if no $y_{kj} < 0$, there is no upper bound to Δc_{Bk}.

To find the change in the value of the objective function.

The value of the objective function for the price vector c is given by

$$Z = c_B \cdot x_B = \sum_{i=1}^{m} c_{Bi} \cdot x_{Bi}.$$

When $c_{Bk} \in c_B$ is changed to $c_{Bk} + \Delta c_{Bk}$, if Z^* is the value of the objective function then

$$Z^* = \sum_{\substack{i=1 \\ i \neq k}}^{m} c_{Bi} \cdot x_{Bi} + (c_{Bk} + \Delta c_{Bk}) x_{Bk}$$

$$= \sum_{i=1}^{m} c_{Bi} . x_{Bi} + x_{Bk} . \Delta c_{Bk} = Z + x_{Bk} \Delta c_{Bk}.$$

Hence if Δc_{Bk} *(change in* c_{Bk}*) satisfies (3), the solution* $\mathbf{x}_B$ *will remain optimal and the value of the objective function will be improved by an amount* $x_{Bk}. \Delta c_{Bk}$ *where* x_{Bk} *is the basic variable corresponding to* c_{Bk}.

ADDITION OF A NEW VARIABLE TO THE PROBLEM

If a new variable is introduced in a L.P.P. whose optimal solution has been obtained, then the solution of the problem will remain feasible. Addition of an extra variable x_{n+1} to the problem will introduce an extra column say α_{n+1} to the coefficient matrix A and an extra cost c_{n+1} will be introduced in the price vector c. Thus the addition of this extra variable may affect the optimality of the problem.

For the same basis B the solution $\mathbf{x}_B$ of the original problem will remain optimal (maxima) if

$$c_{n+1} - Z_{n+1} \leq 0$$

$$\Rightarrow \quad c_{n+1} - c_B B^{-1} \alpha_{n+1} \leq 0. \qquad \text{...(1)}$$

i.e., if (1) is satisfied then the new variable becomes just like a non-basis variable having zero value.

If $c_{n+1} - Z_{n+1} > 0$, then the solution $\mathbf{x}_B$ is no more optimal for the new problem and can be improved by introducing α_{n+1} in the basis. Here we can start with the last simplex table giving the optimal feasible solution of the original problem by introducing one more column corresponding to the variable x_{n+1}.

VARIATION IN THE REQUIREMENT VECTOR b

We know that the condition of optimality for the B.F.S. of a L.P.P. is $\Delta_j = c_j - Z_j \leq 0$.

Since Δ_j does not involve any of b_i if any component b_i of the requirement vector $b = [b_1, b_2, ..., b_m]$ is changed will not affect the conditions of optimality. Hence if any component b_i is changed to $b_i + \Delta b_i$ then the new solution thus obtained will remain optimal. But $x_B = B^{-1} b$ depends on b, therefore any change in b may affect the feasibility of the optimal solution *i.e.*, the optimal solution obtained by changing b may or may not be feasible.

Thus, a change in b_i must be of the magnitude vector b be changed to $b_j + \Delta b_j$, therefore if the new requirement vector is b*, then

$b^* = [b_1, b_2, \dots, b_l + \Delta b_l, \dots, b_m]$.

If x_B^* is the solution of the new L.P.P. obtained by changing b_l to $b_t + Db_l$, then

$x_B^* = B^{-1} b^*$ where B is the optimal basis.

Let $B^{-1} = (\beta_1, \beta_2, \dots, \beta_m)$

$$= \begin{bmatrix} \beta_{11} & \beta_{12} & \cdots & \beta_{1t} & \cdots & \beta_{lm} \\ \beta_{21} & \beta_{22} & \cdots & \beta_{2t} & \cdots & \beta_{2m} \\ \vdots & \vdots & & \vdots & & \vdots \\ \beta_{il} & \beta_{i2} & \cdots & \beta_{il} & \cdots & \beta_{im} \\ \vdots & \vdots & & \vdots & & \vdots \\ \beta_{ml} & \beta_{m2} & \cdots & \beta_{ml} & \cdots & \beta_{mm} \end{bmatrix}$$

l-th component

Since $b^* = [b_1, b_2, \dots, b_l + \Delta b_l, \dots, b_m]$

$= [b_1 + 0, b_2 + 0, \dots, b_l + \Delta\, b_l, \dots, b_m + 0]$

$= [b_1, b_2, \dots, b_l, \dots, b_m] + [0, 0, \dots, \Delta\, b_l, \dots, 0]$

$= b + [0, 0, \dots, \Delta\, b_l, \dots, 0]$

$\therefore\ x_B^* = B^{-1}. b^*$

$= B^{-1}. \{b + [0, 0, \dots, \Delta b_i, \dots, 0]\}$

$= B^{-1}\, b + B^{-1}. [0, 0, \dots, \Delta b_i, \dots, 0]$

l-th component

$= x_B + [\beta_1, \beta_2, \dots, \beta_l, \dots, \beta_m]. [0, 0, \dots, \Delta b_l, \dots, 0]$

$= x_B + \beta_l. \Delta b_l$

$= [x_{B1}, x_{B2}, \dots, x_{Bl}, \dots, x_{Bm}] + [\beta_{1l}, \beta_{2l}, \dots, \beta_{ll}, \dots, \beta_{ml}] . \Delta b_l.$

$= [x_{B1} + \beta_{1l}\, \Delta b_l, \dots, x_{Bl} + \beta_{ll}\, \Delta b_l, \dots, x_{Bm} + \beta_{ml}. \Delta b_l]$ ...(1)

Now if the solution x_B^* is feasible, then

$x_{Bi} + \beta_{il}\, \Delta b_l \geq 0$ for all $i = 1, 2, \dots, m$

$\Rightarrow\ \beta_{il}\, \Delta b_l \geq -x_{Bi}$

$\therefore\ \Delta b_l \geq -\dfrac{x_{Bi}}{\beta_{it}}$, for $\beta_{il} > 0$

and $\Delta b_l \leq -\dfrac{x_{Bi}}{\beta_{il}}$. for $\beta_{il} < 0$.

Hence the range of Δ_l so that the optimal solution x_B^* also remains feasible is given by

$$\underset{\beta_{il}>0}{\text{Max}}\left[\frac{x_{Bi}}{\beta_{il}}\right] \le \Delta b_l \le \underset{\beta_{il}<0}{\text{Min.}}\left[-\frac{x_{Bi}}{\beta_{il}}\right]$$

and the new value of the optimal solution is given by (1).

To find the change in the value of the objective function.

The given value of the objection function for a requirement vector is given by

$$Z = c_B\, x_B = \sum_{i=1}^{m} c_{Bi} \cdot x_{Bi}\,.$$

When b_l is changed to $b_l + \Delta b_l$* if Z* is the value of the objective function, then $Z^* = c_B\, x_B$

$$= (c_{B1}, c_{B2}, \ldots, c_{Bm}).\ [x_{B1} + b_{1l}\, \Delta_{bl}, \ldots, x_{Bm} + b_{ml}\, \Delta b_l]$$

$$= \sum_{i=1}^{m} c_{Bi} \cdot (x_{Bi} + \beta_{il} \cdot \Delta b\ \ )$$

$$= \sum_{i=1}^{m} c_{Bi} \cdot x_{Bi} + \sum_{i=1}^{m} x_{Bi} \cdot \beta_{il} \cdot \Delta b_l$$

$$= Z + \sum_{i=1}^{m} x_{Bi} \cdot \beta_{il} \cdot \Delta b_l$$

Hence if Δb_l (change in b_l) satisfies (2), then the solution x_B given by (1) is also optimal feasible solution. and the value of the objective function is improved by an amount*

$$\sum_{i=1}^{m} x_{Bi} \cdot \beta_{il} \cdot \Delta b_l\,.$$

SOLVED EXAMPLES

Example 1:

A salesman estimate that the following would be the cost on his route visiting the six cities as shown.

To city

		1	*2*	*3*	*4*	*5*	*6*
	1	∞	20	23	27	29	24
	2	21	∞	19	26	31	24
From city	*3*	26	28	∞	15	36	26
	4	25	16	25	∞	23	18
	5	23	20	23	31	∞	10
	6	27	18	12	35	16	∞

The sales man can visit each of the cities once and only once. Determine the optimum sequence he should follow to minimize the total distance travelled. What as the total distance travelled?

Solution:

Following the usual procedure of assignment algorithms, we obtain Table 12.1 showing an optimum solution indicated by encircled zeros.

This table provides the optimum assignment schedule

$1 \rightarrow 3 \rightarrow 4 \rightarrow 2 \rightarrow 1$ and $5 \rightarrow 6 \rightarrow 5$

with distance according to this optimum route being 101.

Table

	1	2	3	4	5	6
1	∞	0	0	7	2	14
2	0	∞	0	10	8	8
3	6	13	∞	0	14	11
4	4	0	6	∞	0	2
5	8	30	10	21	∞	0
6	13	6	0	26	0	∞

This table does not provide the solution to the travelling salesman problem, as it gives $1 \rightarrow 3$, $3 \rightarrow 4$, $4 \rightarrow 2$, $2 \rightarrow 1$, while city 2 is not allowed to follow city 1 unless city 5 and city 6 processed. Now we try to find the next best solution which also satisfies this extra restriction. The next minimum (non-zero) element in the matrix is 2. Therefore, we try to bring 2 into the solution. But the element 2 occurs at two places. We shall consider both the cases separately until the acceptable solution is attained,

	1	2	3	4	5	6
1	∞	0	0	7	(2)	14
2	0	∞	0	10	8	8
3	6	13	∞	0	14	11
4	4	0	6	∞	0	2
5	8	30	10	21	∞	0
6	13	6	0	26	0	∞

We start with the element (1, 5) and make the assignment in this cell instead of zero assignment in the cell (1, 3). After making this assignment we see that no other assignment can be made in the first row and third column, and thus the resulting feasible solution will be

$$1 \to 5,\ 5 \to 6,\ 6 \to 3,$$
$$3 \to 4,\ 4 \to 2,\ 2 \to 1.$$

The selected elements for this solution are encircled in Table 12.2. The cost corresponding to this feasible solution is 2. Again, if we make assignment in the cell (4, 6) instead of (4, 2) then no feasible solution is available in terms of zeros. Hence the best programme is

$$1 \to 5 \to 6 \to 3 \to 4 \to 2 \to 1.$$

The total set-up cost according to this route comes out to be 103.

Solve the travelling salesman problem given by the following data:

$$c_{12} = 20,\ c_{13} = 4,\ c_{14} = 10,$$
$$c_{23} = 5,\ c_{24} = 6,\ c_{25} = 10,$$
$$c_{35} = 6 \text{ and } c_{45} = 20,$$

where $c_{ij} = c_{ij}$

and there is no route between cities i and j if a value for c_{ij} is not shown above.

Given the following matrix of set-up cost, show how to sequence production so as to minimize set-up cost per cycle:

		To				
		A	B	C	D	E
	A	∞	2	5	7	1
	B	6	∞	3	8	2
From	C	8	7	∞	4	7
	D	12	4	6	∞	5
	E	1	3	2	8	∞

A medical representative has to visit five stations A, B, C, D, and E. He does not want to visit any station twice before completing his four of all the stations, and wishes to return to the starting station. Costs of going from one stations to another are given below. Determine the optimal route:

	A	B	C	D	E
A	∞	2	4	7	1
B	5	∞	2	8	2
C	7	6	∞	4	6
D	10	3	5	∞	4
E	1	2	2	8	∞

Example 2(a):

Find an optimal solution to the following L.P. problem

$$\text{Max. } Z = 3x_1 + 5x_2$$

s.t. $x_1 \leq 4,\ x_2 \leq 6$

$$3x_1 + 2x_2 \leq 18 \text{ and } x_1,\ x_2 \geq 0,$$

what happens to this optimal solution if the objective is changed to

$$Z^* = 3x_1 + x_2.$$

Solution:

Introducing the slack variables x_3, x_4 and x_5 the given problem in standard form for simplex method is as follows:

$$\text{Max. } Z = 3x_1 + 5x_2 + 0.x_3 + 0.x_4 + 0.x_5$$

s.t $1.x_1 + x_3 = 4$

$$0.x_1 + 1.x_2 + x_4 = 6$$

$$3x_1 + 2x_2 + x_5 = 18$$

Proceeding as usual we get the following simplex table.

		c_j	3	5	0	0	0	Min. ratio x_B/Y_2
B	c_B	x_B	Y_1	Y_2	Y_3	Y_4	Y_5	
Y_3	0	4	1	0	1	0	0	–
Y_4	0	6	0	(1)	0	1	0	6/1 ←
Y_5	0	18	3	2	0	0	1	18/2
Z = 0		Δ_j	3	5 ↑	0	0 ↓	0	x_B/Y_1
Y_3	0	4	1	0	1	0	0	4/1
Y_2	5	6	0	1	0	1	0	–
Y_5	0	6	(3)	0	0	–2	1	6/3 ←
Z = 30		Δ_j	3 ↑	0	0	–5	0 ↓	
Y_3	0	2	0	0	1	2/3	–1/3	
Y_2	5	6	0	1	0	1	0	
Y_1	3	2	1	0	0	–2/3	1/3	
Z = 36		Δ_j	0	0	0	–3	–1	

and $x_1, x_2, \ldots, x_5 \geq 0$.

Taking $x_1 = 0 = x_2$, we get $x_3 = 4$, $x_4 = 6$,

$x_5 = 18$, which is the starting B.F.S.

From the above table the

$\therefore$ Optimal solution of the given L.P.P. is

$x_1 = 2, x_2 = 6$

and Max. $Z = 36$.

Now the objective function is changed to $Z^* = 3x_1 + x_2$

i.e., c_2 is changed to 1 keeping c_1 fixed.

To find variation in c_2. Here $c_2 \in c_B$, and from last simplex table

$$c_B = (c_{B1}, c_{B2}, c_{B3}) = (0, 5, 3)$$
$$= (c_3, c_2, c_1)$$

$\therefore$ $c_2 = c_{Bk} = c_{B2} = 5$.

$\therefore$ Range of Δc_{B2} is given by

$$\underset{y_{2j} > 0}{\text{Max}} \left[\frac{c_j - Z_j (= \Delta_j)}{y_{2j}} \right] \leq \Delta c_{B2} \leq \underset{y_{2j} < 0}{\text{Min}} \left[\frac{c_j - Z_j (= \Delta_j)}{y_{2j}} \right]$$

for all j not in the basis.

$$\Rightarrow \text{Max.} \left[\frac{\Delta_4}{y_{24}} \right] \leq \Delta c_{B2} < \infty$$

$\Rightarrow$ $-3 \leq \Delta c_{B2} < \infty$

$\therefore$ $5 - 3 \leq c_{B2} < 5 + \infty$ or $2 \leq c_2 < \infty$

$\therefore$ If c_2 is changed to 1, then the optimal solution given in the above table does not remain optimal.

To find the new optimal solution. When Z is changed to Z^*.

If the objective function Z is changed to $Z^* = 3x_1 + x_2$ then c_B in the last simplex becomes (0, 1, 3).

$\therefore$ $\Delta_1 = 0\ \Delta_2 = \Delta_3$,

$\Delta_4 = c_4 - c_B Y_4 = 1$,

$\Delta_5 = c_5 - c_B Y_5 = -1$.

Thus from the last of the above table, we get the new table as follows:

B	c_B	c_j / x_B	3 / Y_1	1 / Y_2	0 / Y_3	0 / Y_4	0 / Y_5	Min. ratio x_B/Y_4
Y_3	0	2	0	0	1	(2/3)	–1/3	2/(2/3) ←
Y_2	1	6	0	1	0	1	0	6/1
Y_1	3	2	1	0	0	–2/3	1/3	
$Z^* = 12$	Δ_j	0	0	0	1 ↓	–1 ↑		
Y_4	0	3	0	0	3/2	1	–1/2	
Y_2	1	3	0	1	–3/2	0	1/2	
Y_1	3	4	1	0	1	0	0	
$Z^* = 15$	Δ_j	0	0	–3/2	0	–1/2		

∴ The solution of the new L.P.P. with changed objective function is $x_1 = 4$, $x_2 = 3$ and Max. $Z^* = 15$.

Example 2(b):

Find an optimal solution to the following L.P.P.

$$\text{Max. } Z = 15x_1 + 45x_2$$

$$\text{s.t.} \quad x_2 \leq 50$$

$$x_1 + 1.6x_2 \leq 240$$

$$5x_1 + 2.9x_2 \leq 162$$

and x_1, x_2 ³ $0.$

If Max. $Z = \sum c_i x_i$ $(i = 1, 2)$ *and* c_2 *is kept fixed at 45, find how much can* c_1 *be changed without affecting the above optimal solution.*

Solution:

Introducing the slack variable x_3, x_4 and x_5, the given problem in standard form for simplex method is as follows:

$$\text{Max. } Z = 15x_1 + 45x_2 + 0.x_3 + 0.x_4 + 0.x_5$$

$$\text{s.t.} \quad 0.x_1 + x_2 + x_3 = 50$$

$$x_1 + 1.6x_2 + x_4 = 240$$

$$5x_1 + 2.0x_2 + x_5 = 162$$

and $x_1, x_2, x_3, x_4, x_5 \geq 0$.

Taking $x_1 = 0 = x_2, x_3 = 50, x_4 = 240, x_5 = 162$, which is the starting B.F.S.

Solving the given L.P.P. by simple method, we get the following simplex table.

B	c_B	c_j / x_B	15 / Y_1	45 / Y_2	0 / Y_3	0 / Y_4	0 / Y_5	Min. ratio x_B/Y_2
Y_3	0	50	0	(1)	1	0	0	50/1
								←
Y_4	0	240	1	1.6	0	1	0	240/1.6
Y_5	0	162	.5	2	0	0	1	162/2
$Z = c_B x_B = 0$		Δ_j	15	45	0	0	0	x_B/Y_1
				↑	↓			
Y_2	45	50	0	1	1	0	0	–
Y_4	0	160	1	0	–1.6	1	0	160/1
Y_5	0	62	(.5)	0	–2	0	1	26/.5
								←
$Z = 2250$	Δ_j	15	0	–45	0	1	x_B/Y_3	
		↑				↓		
Y_2	45	50	0	1	1	0	0	50/1
Y_4	0	36	0	0	(2.4)	1	–2	36/2.4
								←
Y_1	15	124	1	0	–4	0	2	–
$Z = 4110$	Δ_j	0	0	15	0	–30		
				↑	↓			
Y_2	45	35	0	1	0	–5/12	5/6	
Y_3	0	15	0	0	1	5/12	–5/6	
Y_1	15	184	1	0	0	5/3	–4/3	
$Z = 4335$	Δ_j	0	0	0	–25/4	–35/2		

From the table we set the optimal solution of the given L.P.P is

$x_1 = 184, x_2 = 35$, Max. $Z = 4335$.

To find variation in c_1. Here $c_1 \in c_B$.

From the above simplex table, we have

$c_B = (c_{B1}, c_{B2}, c_{B3}) = (45, 0, 15) = (c_2, c_3, c_1)$

$\therefore$ $c_1 = c_{Bk} = c_{B3} = 15.$

$\therefore$ We have to find the range of variation in $c_1 \in c_B$.

Here we want to find the range Δc_{B3} (change in c_{B3}), which is given by (3), 2.

$$\underset{y_{kj}>0}{\text{Max}}\left[\frac{c_j - Z_j(=\Delta_j)}{y_{kj}}\right] \le \Delta c_{Bk} \le \underset{y_{kj}<0}{\text{Min}}\left[\frac{c_j - Z_j(=\Delta_j)}{y_{kj}}\right]$$

Here k = 3.

Here $y_{kj} = y_{3j} = (1, 0, 0, -4/3)$, out of which y_{31}, y_{32}, y_{33} are not to be taken j = 1, 2, 3 correspond to $\alpha_1, \alpha_2, \alpha_3$ which are in the basis. Thus here we shall consider y_{34} and y_{35} only.

Now $y_{34} = 5/3 > 0$

and $y_{35} = -4/3 < 0.$

$\therefore$ Δc_{B3} is given by

$$\text{Max.}\left[\frac{\Delta_4}{y_{34}}\right] \le \Delta c_{B3} \le \text{Min.}\left[\frac{\Delta_5}{y_{35}}\right]$$

or $$\text{Max.}\left[\frac{-25/4}{5/3}\right] \le \Delta c_{B3} \le \text{Min.}\left[\frac{-35/2}{-4/3}\right]$$

or $-15/4 \le \Delta c_{B3} \le 105/8.$

$\therefore$ Range of variation of $c_{B3} = c_1$ is given by

$c_{B3} + (-15/4) \le c_1 \le c_{B3} + 105/8$

or $15 + (-15/4) \le c_1 \le 15 + 105/8$

or $45/4 \le c_1 \le 225/8.$

If we take $c_1 = 45/4$, then proceeding as in above table, we do not get the same optimal solution.

$\therefore$ The variation in c_1 without affecting the above optimal solution is given by $45/4 < c_1 \le 225/8$.

Example 3(a):

Given the following L.P.P.

$$Max.\ Z = -x_1 + 2x_2 - x_3$$

s.t. $3x_1 + x_2 - x_3 \le 10$

$$-x_1 + 4x_2 + x_3 \geq 6$$
$$x_2 + x_3 \leq 4$$

and $x_1, x_2, x_3 \geq 0.$

Find the separate range of b_1, b_2 *and* b_3 *(the constants on the right hand sides of the constraints) consistent with the optimal solution.*

Solution:

Introducing the slack variables x_4, x_6, surplus variables x_5 and artificial variables x_7, the given L.P.P. reduces to

$$\text{Max. } Z = -x_1 + 2x_2 - x_3 + 0.\ x_4 + 0.x_5 + 0.x_6 - Mx_7$$

s.t. $3x_1 + x_2 - x_3 + x_4 = 10$

$-x_1 + 4x_2 + x_3 - x_5 + x_7 = 6$

$x_2 + x_3 + x_6 = 4$

and $x_1, \ldots ,x_7 \geq 0.$

Proceeding as usual, the successive tables of simplex method are as follows

B	c_B	c_j / x_B	−1 / Y_1	2 / Y_2	−1 / Y_3	0 / Y_4	0 / Y_5	−M / Y_6	A_1	Min. ratio x_B/Y_2
Y_4	0	10	3	1	−1	1	0	0	0	10/1
A_1	−M	6	−1	(4)	1	0	−1	0	1	6/4 ←
Y_6	0	4	0	1	1	0	0	1	0	4/1
$Z=c_B\,x_B=-6M$		Δ_j	−1 −M	2+4M ↑	−1+M	=	−M	0	0 ↓	x_B/Y_5
Y_4	0	17/2	13/4	0	−5/4	1	1/4	•0		$\frac{17}{2}/\frac{1}{4}$
Y_2	2	3/2	−1/4	1	1/4	0	−1/4	0		$-\frac{5}{2}/\frac{1}{4}$
Y_6	0	5/2	1/4	0	3/4	0	(1/4)	1		←
$Z=c_B\,x_B=3$		Δ_j	−1/2	0	−3/2	0	1/2 ↑	0 ↓		
Y_4	0	6	3	0	−2	1	0	−1		
Y_2	2	4	0	1	1	0	0	−1		
Y_5	0	10	1	0	3	0	1	4		
$Z=c_B\,x_B=8$		Δ_j	−1	0	−3	0	0	−2		

We have Optimal solution is $x_1 = 0$, $x_2 = 4$ $x_3 = 0$, Max Z = 8.

Here B = $(\alpha_4, \alpha_2, \alpha_5)$, b = (10, 6, 4)

$$x_B = (x_4, x_2, x_5) = (x_{B1}, x_{B2}, x_{B3}) = (6, 4, 10),$$

$$\therefore \quad B^{-1} = (\beta_1, \beta_2, \beta_3) = \begin{bmatrix} 1 & 0 & -1 \\ 0 & 1 & 1 \\ 0 & 0 & 4 \end{bmatrix}$$

To find variation in b_1. After changing b_1 to $b_1 + \Delta b_1$, the new requirement vector is given by

$$b_B^* = (10 + \Delta b_1, 6, 4)$$

From relation (2), 3 the range of Δb_1 consistent with the optimal solution is given by

$$\underset{\beta_{i1} > 0}{\text{Max.}} \left[\frac{-x_{Bi}}{\beta_{i1}}\right] \le \Delta b_1 \le \underset{\beta_{i1} < 0}{\text{Min.}} \left[\frac{-x_{Bi}}{\beta_{i1}}\right]$$

Here $b_{11} = 1 > 0$. Since no $b_{i1} < 0$,

$\therefore$ there is no upper bound to Δb_1.

$$\therefore \quad \text{We have Max.} \left[-\frac{x_{B1}}{\beta_{11}}\right] \le \Delta b_1 < \infty.$$

$$\Rightarrow \quad -\frac{6}{1} \le \Delta b_1 < \infty.$$

$\therefore$ Range for b_1 is $-6 + 10 \le b_1 < 10 + \infty$

$$\Rightarrow \quad 4 \le b_1 < \infty$$

To find variation in b_2.

After changing b_2 to $b_2 + \Delta b_2$, the new requirement vector is

$$b_B^{**} = (10, 6 + \Delta b_2, 4).$$

From relation (2), 3 the range of Δb_2 consistent with the optimal solution is given by

$$\underset{\beta_{i2} > 0}{\text{Max}} \left[\frac{-x_{Bi}}{\beta_{i2}}\right] \le \Delta b_2 \le \underset{\beta_{i2} < 0}{\text{Min.}} \left[\frac{-x_{Bi}}{\beta_{i2}}\right]$$

Here $\beta_{22} = 1 > 0$. Since no $\beta_{i2} < 0$. $\therefore$ there is no upper bound to Δb_2.

$$\therefore \quad \text{We have Max.} \left[-\frac{x_{B2}}{\beta_{22}}\right] \le \Delta b_2 < \infty$$

$$\Rightarrow \quad \frac{-4}{1} \le \Delta b_2 < \infty$$

$\therefore$ Range of variation of b_2 is

$$-4 + 6 \le b_2 < \infty + 0$$

$$\Rightarrow \quad 2 \le b_2 < \infty.$$

To find variation in b_3.

After changing b_3 to $b_3 + \Delta b_3$, the new requirement vector is $b_B{}^{***}$ = (10, 6, 4 + Δ_3), from relation (2), 3 the range of Δb_3 consistent with the optimal solution is given by

$$\underset{\beta_{i3} > 0}{\text{Max.}} \left[\frac{-x_{Bi}}{\beta_{i3}}\right] \le \Delta b_3 \le \underset{\beta_{i3} < 0}{\text{Min.}} \left[\frac{-x_{Bi}}{\beta_{i3}}\right]$$

Here $\beta_{23} = 1$, $\beta_{33} = 4 > 0$ and $\beta_{13} = -1 < 0$.

$$\therefore \quad \text{Max.} \left[\frac{-x_{B2}}{\beta_{23}}, -\frac{x_{B3}}{\beta_{33}}\right] \le \Delta b_3 \le \text{Min.} \left[-\frac{x_{B1}}{\beta_{13}}\right]$$

$$\Rightarrow \quad \text{Max.} \left\{-\frac{4}{1}, -\frac{10}{4}\right\} \le \Delta b_3 \le \text{Min.} \left\{\frac{-6}{-1}\right\}$$

$$\Rightarrow \quad -\frac{5}{2} \le \Delta b_3 \le 6$$

$\therefore$ Range of variation of b_3 is

$$4 - \frac{5}{2} \le b_3 \le 4 + 6$$

$$\Rightarrow \quad \frac{3}{2} \le b_3 \le 10.$$

Example 3(b):

Solve the following L.P.P.

Max. $Z = 3x_1 + 5x_2$

s.t. $x_1 + x_3 = 4$

$3x_1 + 2x_2 + x_4 = 18$

and $x_1, x_2, x_3, x_4 \ge 0$.

If a new variable x_5 is introduced in the above L.P.P. with price 7, then we have the following problem:

Max. $Z' = 3x_1 + 5x_2 + 7x_5$

s.t. $x_1 + x_3 + x_5 = 4$

$$3x_1 + 2x_2 + x_4 + 2x_5 = 18$$

and $x_1, x_2, x_3, x_4, x_5 \geq 0.$

Find the solution of the new L.P.P.

Solution:

Proceeding as usual the successive simplex tables for the original L.P.P. are as follows:

Simplex Table 1

		c_j	3	5	0	0	Mini Ratio
B	c_B	x_B	Y_1	Y_2	Y_3	Y_4	x_B/Y_2
Y_3	0	4	1	0	1	0	–
Y_4	0	18	3	(2)	0	1	18/2 ←
$Z=c_B.x_B=0$		Δ_j	3	5 ↑	0	0 ↓	
Y_3	0	4	1	0	1	0	
Y_2	5	9	3/2	1	0	1/2	
$Z=c_B.x_B=45$		Δ_j	–9/2	0	0	–5/2	

∴ Optimal solution of the given L.P.P. is

$x_1 = 0,\ x_2 = 9,\ x_3 = 4,$

$x_4 = 0$, Max. $Z = 45$.

From the final table the solution of the dual of the given L.P.P. is

$w_1 = 0,\ w_2 = 5/2,$

Min. $Z_D = 45$.

Revised L.P.P. The dual of the new L.P.P. is same as that of the original L.P.P. with one more constraint $w_1 + 2w_2 \geq 7$ which corresponds to the new variable x_5.

The optimal solution of the dual of the original L.P.P. is $w_1 = 0$, $w_2 = 5/2$ which does not satisfy this constraint. Thus we see that the optimal solution of the dual of the given L.P.P. is not optimal for the revised problem and can be improved by introducing $\alpha_5 = [1, 2]$, (column of A corresponding to the new variable introduced) in the basis.

Thus, consider one more column Y_5 in the above table.

Since $B^{-1} = \begin{bmatrix} 1 & 0 \\ 0 & 1/2 \end{bmatrix}$

$\therefore \quad Y_5 = B^{-1}\alpha_5 = \begin{bmatrix} 1 & 0 \\ 0 & 1/2 \end{bmatrix} \cdot \begin{bmatrix} 1 \\ 2 \end{bmatrix} = \begin{bmatrix} 1 \\ 1 \end{bmatrix}$

$\Delta_5 = c_5 - c_B Y_5 = 7 - (0, 5), (1, 1) = 2.$

We get the following simplex table for revised L.P.P. as

Simplex Table 2

B	c_B	c_j / x_B	3 / Y_1	5 / Y_2	0 / Y_3	0 / Y_4	7 / Y_5	Mini Ratio x_B/Y_5
Y_3	0	4	1	0	1	0	(1)	4/1 ←
Y_2	5	9	3/2	1	0	1/2	1	9/1
$Z'=c_B x_B=45$		Δ_j	–9/2	0	0 ↓	–5/2 ↑	2	
Y_5	7	4	1	0	1	0	1	
Y_2	5	5	1/2	1	–1	1/2	0	
$Z'=c_B x_B=53$		Δ_j	–13/2	0	–2	–5/2	0	

$\therefore$ Optimal solution of revised L.P.P. is

$$x_1 = 0, x_2 = 5, x_3 = 0,$$
$$x_4 = 0, x_5 = 4.$$

and Max. Z' = 53.

Example 4(a):

Consider the following table which presents an optimal solution to some linear programming problem.

Table

B	c_B	c_j / x_B	2 / Y_1	4 / Y_2	1 / Y_3	3 / Y_4	2 / Y_5	0 / Y_6	0 / Y_7	0 / Y_8
Y_1	2	3	1	0	0	–1	0	0.5	0.2	–1
Y_2	4	1	0	1	0	2	1	–1	0	0.5
Y_3	1	7	0	0	1	–1	–2	5	–0.3	2
$Z = c_B x_B = 17$		Δ_j	0	0	0	–2	0	–2	–0.1	–2

If the additional constraint

$$2x_1 + 3x_2 - x_3 + 2x_4 - 4x_5 \leq 5$$

were annexed to the system, would there be any change in the optimal solution? Justify your answer.

Solution:

From the table the optimal solution of the given L.P. problem is

$x_1 = 3, x_2 = 1, x_3 = 7,$

$x_4 = 0 = x_5 = x_6 = x_7 = x_8$

which also satisfies the new additional constraint

$2x_1 + 3x_2 - x_3 + 2x_4 - 4x_5 \leq 5.$

Thus the optimal solution of the given problem will not be changed if we introduce the above constraint to it. Hence, the additional constraint is redundant and the optimal sol. of the given L.P.P. is also the optimal solution of the new L.P.P.

Example 4(b):

Given the matrix of setup costs, show how to sequence the production so as to minimize the setup cost per cycle.

To

From		A_1	A_2	A_3	A_4	A_5
	A_1	∞	2	5	7	1
	A_2	6	∞	3	8	2
From	A_3	8	7	∞	4	7
	A_4	12	4	6	∞	5
	A_5	1	3	2	8	∞

Solution:

Consider the problem as an assignment. Applying the assignment technique, we get the following matrix, showing a solution in terms of marked '□' zero:

To

From		A_1	A_2	A_3	A_4	A_5
	A_1	∞	1	3	6	(0)
	A_2	4	∞	(0)	6	⊗
From	A_3	4	3	∞	(0)	3
	A_4	8	(0)	1	∞	1
	A_5	(0)	2	⊗	7	∞

The solution to the assignment problem given by above matrix is

$$A_1 \to A_2, A_5 \to A_1, A_2 \to A_3, A_3 \to A_4, A_4 \to A_2.$$

This solution indicates to produce the products A_1, then A_5 and then again A_1, without producing the products A_2, A_3 and A_4, which violates the additional restriction of prodding each product once and only once before returning to the first product. So this is not a solution of the travelling salesman problem.

Now we try to find the next best solution which also satisfies the additional restriction. The next minimum element (non-zero) in the matrix is 1. We try to bring 1 in the solution. The cost 1 also occurs at three places.

Start by making unity-assignment in the cell (1, 2) instead of zero assignment in the cell (1, 5). Then no other assignment can be made in the first row and the second column. The best solution of the problem lies in the marked '□' elements as shown in the table.

		A_1	A_2	A_3	A_4	A_5
	A_1	∞	(1)	3	6	⊗
	A_2	4	∞	(0)	6	⊗
From	A_3	4	3	∞	(0)	3
	A_4	8	⊗	1	∞	(1)
	A_5	(0)	2	⊗	7	∞

Thus, the required solution of the problem is

$$A_1 \to A_2 \to A_3 \to A_4 \to A_5 \to A_1.$$

For this solution, the cost in the reduced matrix is 2.

On the other hand if we select the element 1 in the cell (4, 3) in the solution, then no feasible solution is available in terms of zero or for which the reduced matrix gives the minimum cost less than 2.

Hence, the most suitable sequence is $A_1 \to A_2 \to A_3 \to A_4 \to A_5 \to A_1$.

The minimum setup cost = 2 + 3 + 4 + 5 + 1 = 15.

Example 4(c):

Solve the following LP problem

Maximize $Z = -x_1 + 3x_3 - 2x_3$

$$3x_1 - x_2 + 2x_3 \le 7$$

$$-2x_1 + 4x_4 \le 12$$

$$-4x_1 + 3x_2 + 8x_3 \le 10$$

and $x_1, x_2, x_3 \geq 0.$

Discuss the effect of the following changes in the optimal solution.

(a) *Determine the range for discrete changes in the coefficients* a_{13} *and* a_{23} *consistent with the optimal solution of the given LP problem.*

(b) *'*x_1*'-column in the problem is changed from* $[3,-2,-4]^T$ *to* $[3, 2, -4]^T$*.*

(c) *'*x_3*'-column in the problem is changed from* $[2, 0, 8]^T$ *to* $[3, 1, 6]^T$*.*

Solution:

The given LP problem in its standard form can be stated as follows:

Maximize $Z = -x_1 + 3x_2 - 2x_3 + 0.s_1 + 0.s_2 + 0.s_3$

subject to the constraints

$$3x_1 - x_2 + 2x_3 + s_1 = 7$$
$$-2x_1 + 4x_2 + s_2 = 12$$
$$-4x_1 + 3x_2 + 8x_3 + s_3 = 10$$

and $x_1, x_2, x_3, s_1, s_2, s_3 \geq 0$

Applying the simplex (Big-M) method, the optimal solution so obtained is shown in Table 1.

Table 1 : Optimal Solution

		$c_j \rightarrow$	*−1*	*3*	*−2*	*0*	*0*	*0*
C_B	***Variables in Basis***	***Solution Values***	x_1	x_2	x_3	s_1	s_2	s_3
	B	$b\ (= x_B)$						
−1	x_1	4	1	0	4/5	2/5	1/10	0
3	x_2	5	0	1	2/4	1/5	3/10	0
0	s_3	11	0	0	10	1	−1/2	1
Z = 11		z_j	−1	3	2/5	1/5	4/5	0
		$c_j - z_j$	0	0	−12/5	−1/5	−4/5	0

The optimal basic feasible solution shown in Table 2 is:

$$x_1 = 4,\ x_2 = 5,\ x_3 = 0$$

and Max Z = 11.

In Table 3.2, the inverse of basis matrix, B is

$$B^{-1} = \begin{bmatrix} 2/5 & 1/10 & 0 \\ 1/5 & 3/10 & 0 \\ 1 & -1/2 & 1 \end{bmatrix} = [\beta_1, \beta_2, \beta_3]$$

Thus, we have $c_B\beta_1 = [-1, 3, 0]\ [2/5, 1/5, 1]^T$
$= -1(2/5) + 3(1/5) + 0(1) = 1/5$

$$c_B\beta_2 = -1(1/10) + 3(3/10) + 0(-1/2) = 8/10$$

$$c_B\beta_2 = -1(0) + 3(0) + 0(1) = 0.$$

Since variables x_1, x_2 and x_3 are in the basis (see column B of Table 3.1), therefore any discrete change in coefficients, belonging to any of these column vectors may affect both feasibility as well as optimality of the original optimal basic feasible solution, whereas any discrete change in the non-basic variables (i.e. x_5, s_1 and s_2) column vectors may affect only optimality condition.

(i) Ranges for discrete change in coefficients a_{13} and a_{23} in the x_3-column vector of Table 3.2 are computed as:

$$\text{Max}\left\{\frac{c_3 - z_3}{c_B\,\beta_1}\right\} = \text{Max}\left\{\frac{-12/5}{1/5}\right\} \le \Delta a_{13} \text{ or } \Delta a_{13} \ge -12$$

and $$\text{Max}\left\{\frac{c_3 - z_3}{c_B\,\beta_2}\right\} = \text{Max}\left\{\frac{-12/5}{8/10}\right\} \le \Delta a_{23} \text{ or } \Delta a_{23} \ge -3$$

(ii) To measure the change Δa_{21} in the coefficient a_{21} (= 2) in the first column of variable x_1 in the second constraint of the original set of constraints, we need to check both feasibility as well as optimality conditions because variable x_1 is the basic variable as shown in Table 3.2.

(a) *Feasibility condition.* For i = 2 (constraint), p = 1 (column), and k = 2, 3 (columns of B^{-1}), we have

For k = 2 $x_{Bk}\,\beta_{pi} - x_{Bp}\,\beta_{12} = x_{B2}\,\beta_{12} - x_{B1}\,\beta_{22}$
$= 5(1/10) - 4(3/10) = -7/10$

For k = 3 $x_{B3}\,\beta_{12} - x_{B1}\,\beta_{32} = 11(1/10) - 4\,(-12) = 32/10$

Hence, the range to maintain feasibility of the existing optimal solution is

$$\frac{-5}{\frac{31}{10}} \le \Delta a_{12} \le \frac{-5}{\frac{-7}{10}}$$

$$2 - (50/31) \le a_{21} \le 2 + (50/7) \text{ or } -12/31 \le a_{21} \le 64/7$$

(b) Optimality condition

$$(c_3 - z_3)\beta_{12} - y_{13}c_B\beta_2 = -\frac{12}{5}\left(\frac{1}{10}\right) - \frac{4}{5}\left(\frac{8}{10}\right) = -\frac{44}{50}$$

$$(c_4 - z_4)\beta_{12} - y_{14}c_B\beta_2 = -\frac{1}{5}\left(\frac{1}{10}\right) - \frac{2}{5}\left(\frac{8}{10}\right) = -\frac{17}{50}$$

$$(c_5 - z_5)\beta_{12} - y_{14}c_B\beta_2 = -\frac{4}{5}\left(\frac{1}{10}\right) - \frac{1}{10}\left(\frac{8}{10}\right) = -\frac{16}{100}$$

Hence, the range to maintain optimality of the existing optimal solution is

$$-\infty \le \Delta a_{12} \le \text{Min}\left\{\frac{-12/5}{-44/50}; \frac{-1/5}{-17/50}; \frac{-4/5}{-16/100}\right\}$$

$$-\infty \le \Delta a_{21} \le 10/17 \text{ or } -\infty \le a_{21} \le 44/17$$

(c) Suppose column vector 33 (x_3-column in Table 3.2) of original LP model is changed from $[2, 0, 8]^T$ to $[3, 1, 6]^T$. Then new value of $c_3 - z^*_3$ for this column is:

$$B^{-1}a_3^* \begin{bmatrix} 2/5 & 1/10 & 0 \\ 1/5 & 3/10 & 0 \\ 1 & -1/2 & 1 \end{bmatrix} \begin{bmatrix} 3 \\ 1 \\ 0 \end{bmatrix} = \begin{bmatrix} 13/10 \\ 9/10 \\ 17/2 \end{bmatrix}$$

$$c_3 - z^*_3 = c_3\, B^{-1}\, a^*_3 = -2 - [-1, 3, 0] \begin{bmatrix} 13/10 \\ 9/10 \\ 17/2 \end{bmatrix} = -\frac{34}{10}$$

Since all entries in the x_3-column are non-zero and so we replace column Xg with new entries in Table 6.10 and proceed to get new optimal solution.

Addition of a New Variable (Column)

Let an extra variable x_{n-1} with coefficient c_{n+1} be added in the system of original constraint. Ax = B, $x \ge 0$. Thus, it creates an extra column a_{n+1} the matrix A ofcoeffcients. To see the impact of this addition on the current optimal solution, we compute

$$V_{n+1} = B^{-1}\, a_{n+1}$$

and $c_{n+1} - z_{n+1} = c_{n+1} - c_B y_{n+1}$

Two cases of the maximization LP model may arise:

1. If $c_{n+1} - z_{n+1} \le 0$, then $x_B = 0$,.and hence current solution remains optimal.
2. If $c_{n+1} - z_{n+1} > 0$, then the current optimal solution can be improved by introducing a new column a_{n+1} into the basis to find the new optimal solution.

Addition of a New Constraint (Row)

After solving an LP model, the decision-maker may recall that a particular resource constraint was overlooked in the model formulation or perhaps he may desire to know the effect of adding a new resource to enhance the objective function value. Addition of a constraint in the existing constraints will cause a simultaneous change in the objective function coefficients (c_j), as well as coefficients a_{ij} of a corresponding non-basic variable. Thus, it will affect only the optimality of the problem. This means that the new variable should enter into the basis only if it improves the value of the objective function.

Suppose that a new constraint

$$a_{m+1,\ 1},\ x_1 + a_{m+1,2}\ x_2 = ...; + a_{m+1,n}\ x_n \le b_{m+1}$$

is added to the system of original constraints $Ax = b$, $x \ge 0$, where b_{m+1}, is positive, zero or negative. Then, two cases may arise.

1. The optimal solution (x_B) of the original problem satisfies the new constraint. If it is so, the solution remains feasible as well as optimal, because the new constraint either reduces or leaves the feasible region of the given LP problem unchanged.
2. The optimal solution (x_B) of the original problem does not satisfy the new constraint. Then the optimal solution to the modified LP problem is re-obtained. Let B be the basis matrix for the original problem and B_1 be the basis matrix for the new problem with m + 1 constraints. That is, matrix B_1 of order (m + 1) is given by

$$B_1 = \begin{bmatrix} B & 0 \\ \alpha & 1 \end{bmatrix}$$

where the second column of B_1 corresponds to slack, surplus or artificial variables added to the new constraint and $\alpha = (a_{m+1,1}, a_{m+1,2}, \ldots a_{m+1,n})$ a is a row vector containing the coefficients in the new constraint and corresponds to variables in the optimal basis. In order to prove that the new solution

$$x_B^* = \begin{bmatrix} X_B \\ s \end{bmatrix} \quad s = \text{slack variable}$$

is a basic feasible solution to the new LP problem, we shall compute the inverse of B_1 using partitioned methods as given below:

$$x_1^{-1} = \begin{bmatrix} B^{-1} & 0 \\ -\alpha B^{-1} & 1 \end{bmatrix}$$

Since each column vector in the new LP problem is given by $a_j^* = (a_j, a_{m+1,j})$, the new columns y_j^*. are given by

$$y_j^* B^{-1} a_j^* = \begin{bmatrix} B^{-1} & 0 \\ -\alpha B^{-1} & 1 \end{bmatrix} \begin{bmatrix} a_j \\ a_{m+1,j} \end{bmatrix}$$

$$= \begin{bmatrix} B^{-1} a_j \\ -\alpha B^{-1} a_j + a_{m+1,j} \end{bmatrix} = \begin{bmatrix} y_j \\ a_{m+1,j} - \alpha_j \end{bmatrix}$$

The entries in the $c_j - z^*_j$ row of the simplex table for any non-basic variable x. in the new problem are computed as follows:

$$c_j - z_j^* = c_j - z_j^* y_j^* = c_j - [c_B, c_{Bm+1}] \begin{bmatrix} y_j \\ a_{m+1,j} - \alpha y_j \end{bmatrix}$$

$$= c_j - (c_B y_j + c_{Bm+1}\, a_{m+1} - c_{Bm+1}\, \alpha_{yj})$$

where cB_{m+1}, is the coefficient associated with the new variable introduced in the basis of the new LP problem.

If the new variable introduced in the basis of the new LP problem is slack or surplus variable, then $c_{Bm+1} = 0$. Hence, we have

$$c_j - z^*_j = c_j - c_B y_j = c_j - z_j$$

That is, the entries in the $c_j - z_j$ row are the same for the initial and new LP problem. The value of objective function is given by

$$Z^* = [c_B, 0] \begin{bmatrix} x_B \\ s \end{bmatrix} = c_B x_B = Z$$

This shows that the optimal simplex table of the original problem remains unchanged even after adding the new constraint. However, the slack or surplus variable appears with negative value because the optimal solution of the original problem does not satisfy the new constraints. Thus, the dual simplex method may be used to get an optimal solution.

Remarks : If the new constraint added is an equation and an artificial variable appears in the basis of the new problem, then two cases may arise:

1. If an artificial variable appears in the basis at negative value, a zero cost may be assigned to it. Apply the dual simplex method to obtain an optimal solution.
2. If the artificial variable appears in the basis at a positive value, then – M cost may be assigned to it. Apply the usual simplex method to obtain an optimal solution.

Example 5:

Solve the travelling salesman problem given by the following data:

$c_{12} = 20,\ c_{13} = 4,\ c_{14} = 10,\ c_{23} = 5,\ c_{34} = 6$

$c_{25} = 10,\ c_{35} = 6,\ c_{45} = 20,$ *where* $c_{ij} = c_{ji}$

and there is no route between cities i and j if the value for c_{ij} *is not given above.*

Solution:

Consider the given problem as an assignment problem.

Taking $c_{ij} = \infty$ for i = j, the cost matrix is as follows:

	1	2	3	4	5
1	∞	20	4	10	∞
2	20	∞	5	∞	10
3	4	5	∞	6	6
4	10	∞	6	∞	20
5	∞	10	6	20	∞

If there is no route between cities i and j then we have taken $c_{ij} = \infty$ to avoid the possibility of going from ith station to jth station.

Now we shall solve the problem by usual assignment algorithm. The following tables show the necessary steps for reaching the solution:

∞	15	(0)	4	∞
15	∞	⊗	∞	3
(0)	⊗	∞	⊗	⊗
4	∞	⊗	∞	12
∞	3	⊗	12	∞

∞	12	(0)	1	∞
12	∞	⊗	∞	(0)
(0)	⊗	∞	⊗	⊗
1	∞	⊗	∞	9
∞	(0)	⊗	9	∞

∞	12	(0)	⊗	∞
11	∞	⊗	∞	(0)
⊗	1	∞	(0)	1
(0)	∞	⊗	∞	9
∞	(0)	⊗	8	∞

Hence, the optimum solution of the assignment problem is:

$1 \to 3,\ 3 \to 4,\ 4 \to 1,\ 2 \to 5,\ 5 \to 2.$

But this is not the solution to the travelling salesman problem, as it is not allowed to go from city 4 to 1 without visiting the cities 2 and 5.

We try to find the 'next best' solution which satisfies the additional restriction. The smallest element other than zero is 1. So we try to bring 1 into the solution. Since the element 1 occurs at two places, we shall consider both the cases separately until the acceptable solution is attained.

Make assignment in the cell (3, 2) having the element 1 instead of making zero assignment in the cell (5, 2). After making this assignment we observe that no other assignment can be made in the third row and second column. Consequently make assignment in the cell (5, 4) having element 8, instead of zero assignment. Consequently make assignment in the cell (5, 2).

The new assignment plan is shown in the following table:

To

From		1	2	3	4	5
	1	∞	12	(0)	⊗	∞
	2	11	∞	⊗	∞	(0)
From	3	⊗	(1)	∞	⊗	1
	4	(0)	∞	⊗	∞	9
	5	∞	⊗	∞	(8)	∞

Thus the resulting feasible solution is

$1 \rightarrow 3 \rightarrow 2 \rightarrow 5 \rightarrow 4 \rightarrow 1$ with cost 9.

Again if we make assignment in the cell (3, 5) having the next best element 1 instead of 0 marked in cell (3, 4), then no feasible solution is obtained in terms of zeros or with cost less than 9.

Hence the optimum route is $1 \rightarrow 3 \rightarrow 2 \rightarrow 5 \rightarrow 4 \rightarrow 1$

The corresponding cost = 4 + 5 + 10 + 20 + 10 = 49.

Example 6:

Solve the following L.P.P.

$$Max.\ Z = 10x_1 + 3x_2 + 6x_3 + 5x_4$$

$$s.t.\ x_1 + 2x_2 + x_4 \leq 6$$

$$3x_1 + 2x_3 \leq 5$$

$$x_2 + 4x_3 + 5x_4 \leq 3$$

$$and\ x_1,\ x_2,\ x_3,\ x_4 \geq 0.$$

Compute the limits for a_{11} and a_{23} so that the new solution remains optimal feasible solution.

Solution:

The given L.P.P. in standard form can be written as follows:

$$\text{Max. } Z = 10x_1 + 3x_2 + 6x_3 + 5x_4 + 0.x_5 + 0.x_6 + 0.\ x_7$$

$$\text{s.t. } x_1 + 2x_2 + 0.x_3 + x_4 + x_5 = 6$$

$3x_1 + 0.x_2 + 2.x_3 + 0.x_4 + x_6 = 5$
$0.x_1 + 1x_2 + 4.x_3 + 5.x_4 + x_7 = 3$
and $x_1, x_2, \ldots, x_7 \geq 0$.

Proceeding as usual the final simplex table is as follows:

		c_j	10	3	6	5	0	0	0
B	c_B	x_B	Y_1	Y_2	Y_3	Y_4	Y_5	Y_6	Y_7
Y_2	3	56/27	0	1	–22/27	0	5/9	–5/27	–1/9
Y_1	10	5/3	1	0	2/3	0	0	1/3	0
Y_4	5	5/27	0	0	26/27	1	–1/9	1/27	2/9
$Z'=c_B\, x_B=643/27$		Δ_j	0	0	–82/27	0	–10/9	–80/27	–7/9

From the above table optimal solution of the given problem is given by

$x_1 = 5/3,\ x_2 = 56/27,\ x_3 = 0,\ x_4 = 5/27$ and $Z = 643/27$.

$\therefore\ B = (Y_2, Y_1, Y_4) = (y_1, y_2, y_3),\ c_B=(c_{B1}, c_{B2}, c_{B3})=(3, 10, 5)$

$x_B = (x_{B1}, x_{B2}, x_{B3}) = (56/27, 5/3, 5/27)\ \alpha_5\ \alpha_6\ \alpha_7$

Since the initial (starting) basis matrix was $(Y_5\ Y_6\ Y_7)$:
from above table

$$B^{-1} = (\beta_1, \beta_2, \beta_3) = \begin{bmatrix} 5/9 & 5/27 & 1/9 \\ 0 & 1/3 & 0 \\ -1/9 & 1/27 & 2/9 \end{bmatrix}$$

i.e., $\beta_{11} = 5/9,\ \beta_{21} = 0,\ \beta_{31} = -\ 1/9,$

$\beta_{12} = -\ 5/27,\ \beta_{22} = 1/3,$

$\beta_{32} = 1/27$ and $\beta_{13} = -\ 1/9,$

$\beta_{23} = 0,\ \beta_{33} = 2/9.$

Example 7:

The linear programming problem is

Max $Z = 3x_1 + 5x_2$

s.t. $x_1 + x_2 \leq 1$

$2x_2 + 3x_2 \leq 1$

and $x_1,\ x_2 \geq 0.$

Obtain the variations in cj (j = 1, 2) which are permitted without changing the optimal solution.

Solution:

Proceeding as usual, the *final simplex table* giving the optimal solution of the given L.P.P. is as follows.

		c_j	3	5	0	0
B	c_B	x_B	Y_1	Y_2 (β_2)	Y_3 (β_1)	Y_4
Y_3	0	2/3	1/3	0	1	–1/3
Y_2	5	1/3	2/3	1	0	1/3
$Z = c_B x_B = 5/3$		Δ_j	–1/3	0	0	–5/3

From the above table the optimal solution, we have

$x_1 = 0$, $x_2 = 1/3$, Max. $Z = 5/3$.

Here $c_B = (c_{B1}, c_{B2}) = (0, 5) = (c_3, c_2)$.

To find variation in c_1. Here c_1 does not belong to c_B.

$\therefore$ From (1), 2, the change Δc_1 in c_1, so that the solution remains optimal, is given by

$$\Delta c_1 \leq Z_1 - c_1 = -\Delta_1.$$

$$\Rightarrow \quad \Delta c_1 \leq 1/3.$$

$\therefore$ The range over which c_1 can vary maintaining the optimality of the solution given in above table is

$$-\infty < c_1 \leq c_1 + \Delta c_1$$

$$\Rightarrow \quad -\infty < c_1 \leq 3 + 1/3$$

$$\Rightarrow \quad -\infty < c_1 \leq 10/3.$$

The value of the objective function will not change in this case.

To find variation in c_2. Here $c_2 = c_{Bk} = c_{B2} = 5 \in c_B$

$\therefore$ From (3), 2 the range of Δc_{B2} (change in c_{B2}) is given by

$$\underset{y_{kj} > 0}{\text{Max}} \left[\frac{c_j - Z_j}{y_{kj}} \right] \leq \Delta c_{Bk} \leq \underset{y_{kj} < 0}{\text{Min}} \left[\frac{c_j - Z_j}{y_{kj}} \right].$$

Here k = 2 and $y_{21} = 2/3$, $y_{24} = 1/3 > 0$. Here we cannot consider y_{22} and y_{23} as α_2 and α_3 are in the basis.

Since no $y_{2j} < 0$,

$\therefore$ There is no upper bound to Δc_{B2}.

The change in Δc_{B2} is given by

$$\text{Max.} \left[\frac{\Delta_1}{y_{21}}, \frac{\Delta_4}{y_{24}} \right] \leq \Delta c_{b2} < \infty$$

$$\text{Max.} \left[\frac{-1/3}{2/3}, \frac{-5/3}{1/3}\right] \leq \Delta c_{B2} < \infty$$

$$\Rightarrow \quad -1/2 \leq \Delta c_{B2} < \infty$$

$$\Rightarrow \quad c_{B2} - 1/2 \leq c_2 < c_{B2} + \infty$$

$$\Rightarrow \quad 5 - 1/2 \leq c_2 < 5 + \infty$$

$$\Rightarrow \quad 9/2 \leq c_2 < \infty.$$

EXERCISES

1. A company makes two products: A and B. The production of both products requires processing time' in two departments I and II. The hourly capacity of I and II, unit profits for products: A and B and the processing time requirements in I and II are given in the following table:

Department	*Product*		*Capacity (Hours)*
	A	*B*	
I	1	2	32
II	0	1	8
Unit profit (Rs.)	200	300	

   Now the company is considering the addition of a new product C to its line. Product C requires one hour each of department I and II. What must be product C's unit profit in order to profitably add it to the firm's product line?

2. A pig farmer is attempting to analyse his feeding operation. The minimum daily requirement of the three nutritional elements for the pigs and the number of units of each of these nutritional elements in two feeds is given in the following table:

Required Nutritional Element	*Units of Nutritional Elements (in kg)*		*Minimum Requirement*
	Food 1	*Food 2*	
A	20	30	200
B	40	25	350
C	30	45	430
Cost per kg	5	3	

   (a) Formulate and solve this problem as an LP model.

(b) Assume that the farmer can purchase a third feed at a cost of Rs. 2 per kg, which will provide 35 units units of nutrient A, 30 units of nutrient B and 50 units of nutrient C. Would this change the optimal mix of feeds? If yes, how?

3. Write a short note on sensitivity analysis.

4. Given that the problem: Max. $Z = cx$ such that $Ax = b$, $x \geq 0$, has an optimal solution; can one obtain a linear programming problem which has an unbounded solution changing b along?

5. For the L.P.P. Max. $Z = 5x_1 + 3x_2$

Subject to, $3x_1 + 5x_2 \leq 15$

$5x_1 + 2x_2$ £ 10

and x_1, x_2 ³ 0.

Find an optimal solution. Hence find how far the component c_1 of the vector c of the function $Z = cx$ can be increased without destroying the optimality of the solution.

6. Solve the following L.P.P.

Max. $Z = 3x_1 + 5x_2$

s.t. $x_1 + x_3 = 4$

$3x_1 + 2x_2 + x_4 = 18$

and x_1, x_2, x_3, x_4 ³ 0.

Would there be any change in the optimal solution if the additional constraint (1) x_2 £ 10 or (ii) x_2 £ 6 is added to the above L.P.P. In case the optimal solution change find the new optimal feasible solution.

7. (i) Discuss the effect of discrete changes in the requirements (on the right side of the inequalities) for the following L.P.P.

Max. $Z = 3x_1 + 4x_2 + x_3 + 7x_4$

subject to

$8x_1 + 3x_2 + 4x_3 + x_4 \leq 7$

$2x_1 + 6x_2 + x_3 + 5x_4 \leq 3$

$x_1 + 4x_2 + 5x_3 + 2x_4 \leq 8,$

$x_1\ x_2, x_3, x_4 \geq 0.$

(ii) Discuss the effect of discrete changes in c on the optimality of an optimum basis feasible solution to the above L.P.P.

8. Solve the following L.P.P. and find how much can c_1 be changed without affecting the optimal conditions.

$$\text{Max. } Z = 15x_1 + 45x_2 = c_1 x_1 + c_2 x_2$$
$$\text{s.t. } x_1 + 1.6x_2 \leq 240$$
$$.5x_1 + 2x_2 + x_3 = 162,$$
$$x_2 + x_4 = 50,$$
$$x_1, x_2, x_3, x_4 \geq 0.$$

9. Write a short note on sensitivity analysis.
10. Discuss the role of sensitivity analysis in linear programming. Under what circumstances is it needed, and under what conditions do you think it is not necessary?
11. (a) Explain how a change in a input-output coefficient can affect a problem's optimal solution?
    (b) How can a change in resource availability affect a solution?
12. What do you understand by the term 'sensitivity analysis'? Discuss the effect of (i) variation of c_j, (ii) variation of b_i, and (iii) addition of a new constraint.
13. A company sells two different products: A and B. The selling price and incremental cost information is as follows:

	Product A	*Product B*
Selling price (Rs.)	60	40
Incremental cost (Rs.).	30	10
Incremental profit (Rs.)	30	30

The two products are produced in a common production process and are sold in two different markets. The production process has a capacity of 30,000 labour hours. It takes three hours to produce a unit of A and one hour to produce a unit ofB. The market has been surveyed and company officials feel that the maximum number of units of A that can be sold is 8,000; and that of B is 12,000 units.

(a) Find the optimal product mix.

(b) Suppose maximum number of units of A and B that can be sold is actually 9,000 units and 13,000 units, respectively instead of as given in the problem, what effect does this have on the solution? What is the effect on the profit? What is the shadow price for the constraint on the sales limit for both these products.

(c) Suppose there are 31,000 labour hours available instead of 30,000 as in the base case, what effect does this have on the solution?

What is the effect on the profit? What is the shadow price for the constraint on the number of labour hours.

14. A firm uses three machines in the manufacture of three products. Each unit of product I requires 3 hours on machine 1, 2 hours on .machine 2 and 1 hour on machine 3. Each unit of product II requires 4 hours on machine 1, 1 hour on machine 2 and 3 hours on machine 3. Each unit of product III requires 2 hours on machine 1, 2 hours on machine 2, and 2 hours on machine 3. The contribution margin of the three products is Rs. 30, Rs. 40 and Rs. 35 per unit, respectively. Available for scheduling are 90 hours of machine 1 time, 54 hours of machine 2 time, and 93 hours of machine 3 time.
   (a) What is the optimal production schedule for the firm?
   (b) What is the marginal value of an additional hour of time on machine?
   (c) What is the opportunity cost associated with product I? What interpretation should be given to this opportunity cost?
   (d) Suppose that the contribution margin for product I is increased from Rs. 30 to Rs. 43, would this change the optimal production plan? Give reasons.

15. An organization can produce a particular component for passenger cars, jeeps and trucks. The production of the component requires utilization of sheet metal working and painting facilities, the details of which are given below.

Resource	*Consumption to Produce a Unit for*			*Availability*
	Passenger car	*Jeep*	*Truck*	
Sheet metal working	0.25 hr	1 hr	0.5 hr	12 hrs
Painting	0.5 hr	1 hr	2 hr	30 hrs

The profits that can be earned by the three categories of components, i.e. for passenger cars, jeeps and trucks are Rs. 600, Rs. 1400, and Rs. 1300, respectively.
   (a) Find the optimal product mix for that organization.
   (b) What additional profit would be earned by increasing the availability of:
      (i) Sheet metal working shop by an hour only.
      (ii) Painting shop by an hour only.
   (c) What would be the effect on the profit earned if at least one component for jeep had to be produced?

16. Find an optimal solution to the following L.P.P.

    Max. $Z = 15x_1 + 45x_2$

    s.t. $x_1 + 16x_2 \leq 240$

    $5x_1 + 2x_2 \leq 162$

    $x_2 \leq 50,\ x_1,\ x_2 \geq 0.$

    If max. $Z = \sum c_j x_j$, $j = 1, 2$ and c_2 is fixed at 45, determine how much c_1 can be changed without affecting the above solution.

17. In a LP problem. Max Z = ex, subject to Ax = b; $x \geq 0$, obtain the variation in coefficients c. which are permitted without changing the optimal solution. Conduct sensitivity analysis on c_j's for the following LP problem.

    Max $Z = 3x_1 + 5x_2$

    subject to $x_1 + x_2 \leq 1$

    $2x_1 + 2x_2 \leq 1$

    and $x_1, x_2, \geq 0$

18. Given the LP problem

    Max $Z = x_1 + 2x_2 + x_3$

    subject to $3x_1 + x_2 - x_3 \leq 10$

    $-x_1 + x_2 + x_3 \leq 6$

    $x_2 + x_3 \leq 4$

    and $x_1, x_2, x_3 \geq 10$

    (a) Determine optimal solution to the problem

    (b) Determine the effect of discrete changes in those components of the cost vector which corresponds to the basic variable.

20. Discuss the changes in the coefficients a_{ij} for the given LP problem: Max Z = ex, subject to Ax = b, $x \geq 0$.

21. Given, that the problem: Max Z = c x such that A x = b, $x \geq 0$ has an optimal solution, can one obtain a linear programming problem which has an unbounded solution changing b alone.

22. Consider the LP problem; Max Z = ex; subject to Ax = b and $x \geq 0$, where $c, x^T \in E^n$, $b^T \in E^m$ and A is $m \times n$ coefficients matrix. Determine how much the components of the cost vector c can be changed without affecting the optimal solution of the LP problem.

23. Find the limits of variation of element a_{ik}, so that optimal feasible solution of Ax = b, $x \geq 0$, Max Z = ex remains optimal feasible solution when (i) $a_k \in B$, (ii) $a_k \notin B$.

24. Solve the 'travelling salesman problem' given by the following data:

    $c_{12} = 4$, $c_{13} = 7$, $c_{14} = 3$, $c_{23} = 6$, $c_{24} = 3$ and $c_{34} = 7$, where $c_{ij} = c_{ji}$.

25. Find the optimal solution to the LP problem

    Max $Z = 15x_1 + 45x_2$

    subject to
    $$x_1 + 16x_2 \leq 250$$
    $$5x_1 + 2x_2 \leq 162$$
    $$x_2 \leq 50$$
    and
    $$x_1, x_2 \geq 0$$

    If Max $Z = \Sigma\, c_j x_j, j = 1, 2$ and c_2 is kept fixed at 45, determine how much can c_1 be changed without affecting the optimal solution of the problem.

26. Discuss the effect of discrete changes in the parameter b_1 (i = 1, 2, 3) for the LP problem.

    Max $Z = 3x_1 + 4x_2 + x_3 + x_4$

    subject to
    $$8x_1 + 3x_2 + 4x_3 + x_4 \leq 7$$
    $$2x_1 + 6x_2 + x_3 + 5x_4 \leq 3$$
    $$x_1 + 4x_2 + 5x_3 + 2x_4 \leq 8$$
    and
    $$x_1, x_2, x_3, x_4 \geq 0$$

27. Write a short note on travelling salesman problem.

28. State the travelling salesman problem and formulate it as an assign ment problem.

29. How can the travelling select man problem be solved using assignment algorithm.

30. Given the LP problem

    subject to
    $$3x_1 + x_2 - 4x_3 \leq 10$$
    $$-x_1 + 4x_2 + x_3 \geq 6$$
    $$x_2 + x_3 \leq 4$$
    and
    $$x_1, x_2, x_3 \geq 0.$$

    Determine the range for discrete changes in the resource values $b_1 = 10$ and $b_2 = 6$ of the LP. model so as to maintain optimality of the current solution.

31. Given the LP problem

    Max $Z = 3x_1 + 5x_2$

subject to $3x_1 + 2x_3 \leq 18$

$x_1 \leq 4$

$x_2 \leq 6$

and $x_1, x_2 \geq 0$

Discuss the effect on the optimality of the solution when the objective function is changed to $3x_1 + x_2$.

32. A stainless steel utensil manufacturer makes three types of items. The restrictions, profits and requirements are tabulated below:

Utensil Type	*I*	*II*	*III*
Raw material requirement (kg per unit)	6	3	5
Welding and finishing time (hours per unit)	3	4	5
Profit per unit (Rs.)	3	1	4

If stainless steel (raw material) availability is 25 kg and welding and finishing time available is 20 hours per day, then the optimum product mix problem is expressed as:

Max $Z = 3x_1 + x_2 + 4x_3$

subject to $6x_1 + 3x_2 + 5x_3 \leq 25$ (raw material restriction)

$3x_1 + 4x_2 + 5x_3 \leq 20$ (time restriction)

and $x_1, x_2, x_3 \geq 0$

where x_j $(j = 1, 2, 3)$ is the number of units of the jth type of the item to be produced. Get the optimal simplex table and encircle the appropriate answer to the following questions.

(a) The second type of utensil would change the current optimal basis if its profit per unit is: ≥ 1; ≥ 1.02; ≥ 1.06; ≥ 2; ≥ 3.

(b) The simplex multiplier associated with the machine time restriction of 20 hours is (–3/5). Thus, the multiplier remains unchanged for the upper limit on the machine time availability of 25; 27.5; 35; 42.5; 32.5.

(c) The increase in the objective function for each unit availability of machine time higher than the upper limit indicated in part (b) is ... (show calculations).

(d) The profit of the third type of utensil is Rs. 4 per unit. The lower unit on its profitability such that the current basis is still optimal is 4; 3; 2.5; 2; < 2.

33. Given the LP problem

$\text{Max } Z = x_1 + 2x_2 + x_3$

subject to $3x_1 + x_2 - x_3 \leq 10$

$-x_1 + x_2 + x_3 \leq 6$

$x_2 + x_3 \leq 4$

and $x_1, x_2, x_3 \geq 10.$

(a) Determine the sensitivity limits for unit profits within which the current optimal solution will remain unchanged.

(b) What is the total profit when each table yields a profit of Rs. 180 and each chair yields a profit of Rs. 100.

34. Consider the following LP problem

$\text{Max } Z = 4x_1 + 6x_2$

subject to $x_1 + 2x_2 \leq 8$

$6x_1 + 4x_2 + \leq 24$

and $x_1, x_2 \geq 0$

(a) What is the optimal solution?

(b) If the first constraint is altered as: $x_1 + 3x_2 \leq 8$, does the optimal solution change?

5

Waiting Lines

INTRODUCTION

Waiting lines or queues are omnipresent. Businesses of all types, industries, schools, hospitals, cafeterias, book stores, libraries, banks, post offices, petrol pumps, theatres—all have queuing problems. Further examples of queues, though less apparent are : waiting for a telephone operator 10 answer, a traffic light to change, the morning mail to be delivered and the like, A common situation occurring in everyday life is that of queuing or waiting in a line. Queues (waiting lines) form at bus stops, ticket booths, doctors' clinics, bank counters, traffic lights and so on. Queues are also found in industry, in shops where the machines wait to be repaired; at a tool crib where the mechanics wait to receive tools, in a warehouse where parts wait to be used and in telephone exchanges where incoming calls wait to mature.

The queuing theory or waiting line theory owes its development to A.K. Erlang. He in 1903, look up the problem on congestion of telephone traffic. The difficulty was that during busy periods, telephone operators were unable to handle the calls the moment they were made, resulting in delayed calls. A.K. Erlang directed his first efforts at finding the delay for one operator and later on the results were extended to find the delay for several operators. The field of telephone traffic was further developed by Molins (1927) and Thornton D-Fry (1928). However, it was only after World War- II that this early work was extended to other general problems involving queues or waiting lines.

Waiting line problems arise either because:

(i) there is loo much demand on the facilities so that we say that there is an excess of waiting time or inadequate number of service faciiities.

(ii) there is too less demand, in which case there is loo much idle facility time or too many facilities.

In either case, the problem is to either *schedule arrivals or provide facilities or both* so as obtain an optimum balance between the costs associated with waiting time and idle lime.

Queuing theory can be applied to a wide variety of operational situations where imperfect matching between the customers and service facilities is caused by one's inability to predict accurately the arrival and service times of customers. In particular, it can be used to determine the level of service (either the service rate or the number of service facilities) that balances the following two conflicting costs:

(i) Cost of offering the service

(ii) Cost incurred due to delay in offering service.

The first cost is associated with the service facilities and their operation, and the second represents the cost of customer's waiting time.

Obviously, an increase in the existing service facilities would reduce the customer's waiting time. Conversely, decreasing the level of service should result in long queue(s). This means an increase (decrease) in the level of service increases (decreases) the cost of operating service facilities but decreases (increases) the cost of waiting.

SINGLE-CHANNEL QUEUING THEORY

A single channel queuing problem results from random arrival time and random service time at a single service station. The random arrival time can be described mathematically with a probability distribution. The most common distribution found in queuing problems is Poisson distribution. This is used in single channel queuing problems for random arrivals where the service time is exponentially distributed. The sections ahead give the reader an insight into the true nature of operations research—the difficulties of developing OR models, the need of logical assumptions and the utilization of higher mathematics.

PERFORMANCE MEASURES OF A QUEUING SYSTEM

(Some of the performance measures (operating characteristics) of any queuing system that are of a general interest for the evaluation of the performance of an existing queuing system, and to design a new system in terms of the level of service a customer receives as well as the proper utilization of the service facilities are listed below.

1. Time-Related Questions for The Customers

(a) What is the average (or expected) time an arriving customer has to wait in the queue (denoted by W_q) before being served.

(b) What is the average (or expected) time an arriving customer spends in the system (denoted by W_s) including waiting and service. This data can be used to make economic comparison of alternative queuing systems.

2. Quantitative Questions Related to the Number of Customers

(a) Expected number of customers who are in the queue (queue length) for service, and is denoted by L_q.

(b) Expected number of customers who are in the system either waiting in the queue or being serviced (denoted by L_s). The data can be used for finding the mean customer time spent in the system.

3. Questions Involving Value of Time Both for Customers and Servers

(a) What is the probability that an arriving customer has to wait before being served (denoted by P_w)? It is also called *blocking probability*.

(b) What is the probability that a server is busy at any particular point in time (denoted by ρ)? It is the proportion of the time that a server actually spends with the customer, *i.e.,* the fraction of the time a server is busy.

(c) What is the probability of n customers being in the queuing system when it is in steady state condition? It is denoted by P_n, n = 0, 1....

(d) What is the probability of service denial when an arriving customer cannot enter the system because the queue is full? It is denoted by P_d.

4. Cost-Related Questions

(a) What is the average cost needed to operate the system per unit of time?

(b) How many servers (service centres) are needed to achieve cost effectiveness?

To describe the distribution of these variables, we should-specify its average value, standard deviation and the probability that the variable exceeds a certain value.

Transient-State and Steady-State

When a service system is started it progresses through a number of changes. However, it attains stability after some time. Before the start of the service operations it is very much influenced by the initial conditions (number of customers in the system) and the elapsed time. This period of transition is termed as *transient-state*. However, after sufficient time has passed, the system becomes independent of the initial conditions and of the elapsed time (except under very special conditions) and enters a steady-state condition.

In the development of queuing theory models it is assumed that the system has entered a steady state primarily for two reasons:

(a) System most of the time operates in the steady-state condition, and

(b) Transient case is much more complex.

In this chapter an analysis of the queuing system will be discussed under steady-state conditions.

Let P_n (t) denote the probability that there are n customers in the system at time t. The rate of change in the value P_n (t) with respect to time t is denoted by the derivative of P_n (t) with respect to t, *i.e.*, P_n(.t). In the case of steady-state, we have

$$\lim_{t\to\infty} P_n(t) = P_n \text{ (independent of t)}$$

or $$\lim_{t\to\infty} \frac{d}{dt}\{P_n(t)\} = \frac{d}{dt}(P_n)$$

or $$\lim_{t\to\infty} P_n'(t))) = 0.$$

In some cases when arrival rate of customers in the system is more than the service rate, then a steady-state cannot be reached regardless of the length of the elapsed time.

Line length (or Queue Size) : The total number of customers in the system who are actually waiting in the line and not being serviced.

Queue Length : Queue length may be defined as the line length plus number of customers being served.

Notations : The notations used in the analysis of a queuing system are as follows:

n = number of customers in the system (waiting and in service)

P_n = probability of n customers in the system

λ = average (expected) customer arrival rate or average number of arrivals per unit of time in the queuing system

μ = average (expected) service rate or average number of customers served per unit time at the place of service

$$\frac{\lambda}{\mu} = \rho = \frac{\text{Average service completion time } (1/\mu)}{\text{Average inter - arival time } (1/\lambda)}$$

= traffic intensity or server utilization factor (the expected fiaction of, time for which server is busy).

s = number of service channels (service facilities or servers)

N = maximum number of customers allowed in the system.

L_s = average (expected) number of customers in the system (waiting and in service)

L_q = average (expected) number of customers in the queue (queue length)

L_b = average (expected) length of non-empty queue

W_s = average (expected) waiting time in the system (waiting and in service)

W_q = average (expected) waiting time in the queue

P_w = probability that an arriving customer has to wait.

For achieving a steady-state condition and analytical results to be valid, we must have, $\lambda/\mu < 1$, (*i.e.,* the arrival rate must be less than the service rate). In some queuing models there is no such limitation on the value of ρ. Such a situation arises when the queue length is limited, generally because of space or capacity limitation or in which customers balk when the queue becomes too long

Relationships Among Performance Measures

By definition of various measures of performance (operating characteristic), we have

$$L_s = \sum_{n=0}^{\infty} n\, P_n \text{ and } L_q = \sum_{n=s}^{\infty} (n-s)\, P_n .$$

Some general relationships between the average system characteristics true for all queuing models are as follows:

(i) Expected number of customers in the system is equal to the expected number of customers in queue plus in service.

$$L_s = L_q + \text{Expected number of customers in service}$$

$$= L_q + \frac{\lambda}{\mu}$$

The value of expected number of customers in service, should not be confused with the number of service facilities but it is equal to ρ for all queuing models except finite queue case.

(ii) Expected waiting time of the customer in the system is equal to the average waiting time in queue plus the expected service time

$$W_s = W_q + \frac{1}{\mu}$$

(iii) Expected number of customers served per busy period is given by

$$L_s = \frac{L_s}{P(n \geq s)} = \frac{\mu}{\mu - \lambda}$$

where $P(n \geq s)$ = probability that the system being busy

(iv) Expected length of queue during busy period is given by

$$W_b = \frac{W_q}{P(n \geq s)} = \frac{\lambda}{\mu - \lambda}$$

(v) Expected number of customers in the system is equal to the average number of arrivals per unit of time multiplied by the average time spent by the customer in the system

$$L_s = \lambda W_s \text{ or } W_s = \frac{1}{\lambda} L_s$$

(vi)
$$L_q = \lambda W_q \text{ or } W_q = \frac{1}{\lambda} L_s \text{ or } \frac{L_s}{\mu}$$

For applying formula (v) and (vi) for system with finite queue, instead of using λ, its effective value $\lambda(1 - P_N)$ must be used.

(vii) The probability, P_n of n customers in the queuing system at any time can be used to determine all the basic measures of performance in the following order

$$L_s = \sum_{n=0}^{\infty} nP_n \rightarrow W_s = \frac{L_s}{\lambda} \rightarrow W_q = W_s - \frac{1}{\mu} \rightarrow L_q = \lambda W_q$$

CLASSIFICATION OF QUEUING MODELS AND THEIR SOLUTIONS

Different models in queuing theory are classified by using special (or standard) notations described initially by D.G. Kendall in 1953 in the form (a/b/c). Later A.M. Lee in 1966 added the symbols c/and c to the Kendall notation. Now in the literature of queuing theory the standard format used to describe the main characteristics of parallel queues is as follows:

{(a/b/c) : (d/c)}

where

a = arrivals distribution

b = service time (or departures) distribution

c = number of service channels (servers)

d = maximum number of customers allowed in the system (in queue plus in service)

e = queue (or service) discipline.

Certain descriptive notations are used for the arrival and service times distribution (*i.e.,* to replace notation a and b) as follows:

M = Exponential (or Markovian) inter-arrival times or service-time distribution (or equivalently Poisson or Markovian arrival or departure distribution).

D = Constant or deterministic inter-arrival-time or service-time.

G = Service time (departures) distribution of general type, *i.e.,* no assumption is made about the type of distribution.

GI = Inter-arrival time (arrivals) having a general probability distribution such as normal, uniform or any empirical distribution.

E_k = Erlang-k: distribution of inter-arrival or service time distribution with parameter k (*i.e.,* if k = 1, Erlang is equivalent to exponential and if k =1, Erlang is equivalent to deterministic).

For example, a queuing system in which the number of arrivals is described by a Poisson probability distribution, the service time is described by an exponential distribution, and there is a single server, would be designated by M/M/1.

The Kendall notation now will be used to define the class to which a queuing model belongs. The usefulness of a model for a particular situation is limited by its assumptions.

SOLUTION OF QUEUING MODELS

There are two approaches, namely mathematical and simulation, which are available for solving queuing models. In this chapter, we will discuss only mathematical approach. In the mathematical approach the actual arrival and service time distributions are approximated by one of the known mathematical distributions, *i.e.,* Poisson, exponential, etc., to describe expected value of various operational characteristic of the queuing process. Further, if cost associated with queuing process is also known, then conditions can be determined under which operational characteristics produce optimal results (maximizes the profit or minimizes the total cost).

In this section, we will discuss various queuing situations (models) in which both arrivals and departures (those serviced) take place simultaneously with specific assumptions about the features of queuing systems such as: input source, arrival process, queue configuration, queue discipline and service process. The simultaneous occurrence of arrivals and departures is also called *Birth-and-Death Process*. This process helps in determining the probability distribution of the number of customers in the queuing system at a particular instance of time. The probability distribution so obtained is then used to determine performance measures (operating characteristics) of various types of queuing models.

SINGLE-SERVER QUEUING MODELS

Model I {(M/M/l): (∞/FCFS)} Exponential Service—Unlimited Queue

The derivation of this model is based on certain assumptions about the queuing system:

(i) Exponential distribution of inter-arrival time or Poisson distribution of arrivals.

(ii) Single waiting line with no restriction on length of queue (*i.e.,* infinite capacity) and no balking or reneging.

(iii) Queue discipline is *'first-come, first-served'*

(iv) Single server with exponential distribution of service time.

The following events (possibilities) may occur during a small interval of time, Δt just before time t. It is assumed that the system is in state n (number of customers) at time t.

1. The system is in state n (number of customers) and no arrival and no departure, leaving the total to n customers.
2. This system is in state n + 1 (number of customers) and no arrival and one departure, reducing the total to n customers.
3. The system is in state n – 1 (number of customers) and one arrival and no departure, bringing the total to n customers.

Step 1: Obtain System of Differential Difference Equations

If P_n (t) is the probability of n customers at time t in the system, then the probability that the system will contain n customers at time (t + Δt) can be expressed as the sum of the joint probabilities of the three mutually exclusive and collectively exhaustive cases as mentioned above. That is,

For n ≥ 1 and t ≥ 0,

$$P_n (t + \Delta t) = P_n (t). \text{ Prob (no arrival in } \Delta t \text{ and on departure in } \Delta t)$$

$$+ P_{n+1} (t). \text{ Prob (no arrival in } \Delta t \text{ and one departure in } \Delta t)$$

$$+ P_{n-1} (t). \text{ Prob (one arrival in } \Delta t \text{ and no departure in } \Delta t)$$

$$= P_n (t) \{\text{Prob (no arrival in } \Delta t) \times \text{(no departure in } \Delta t\}$$

$$P_{n+1} (t) \{\text{Prob (no arrival in } \Delta t) \times \text{Prob (one departure in } \Delta t)\}$$

$$+ P_{n-1} (t) \{\text{Prob (one arrival in } \Delta t) \times \text{Prob (no departure in } \Delta\}$$

$$= P_n (t) \{1 - \lambda \Delta t\} \{1 - \mu \Delta t\} + P_{n+1} (t) \{1 - \lambda \Delta t\} \mu \Delta t + P_{n-1} (t) \{\lambda \Delta t\} \{1 - \mu \Delta t\}$$

$$= P_n (t) \{1 - (\lambda + \mu)\Delta t\} + P_{n+1} (t) \mu \Delta t + P_{n-1} (t) \lambda \Delta t$$

+ terms involving $(\Delta t)^2$

Since Δt is very small, therefore terms involving $(\Delta t)^2$ can be neglected. Subtracting P_n (t) from both sides and dividing by Δt, we get

$$\frac{P_n(t+\Delta t) - P_n(t)}{\Delta t} = \lambda P_{n-1}(t) + \mu P_{n+1}(t) - (\lambda + \mu) P_n(t)$$

Taking limit on both sides as $\Delta t \to 0$, then above equation reduces to

$$P'_n(t) = \frac{d}{dt}\{P_n(t)\} = \lambda P_{n-1}(t) + \mu P_{n+1}(t)$$
$$+ (\lambda + \mu) P_n(t);\ n \geq 1 \qquad ...(1)$$

Similarly, it there is no customer in the system at time $(t + \Delta t)$, then there will be no service completion during Δt. Thus for $n = 0$ and $t \geq 0$, we have only two probabilities instead of three. The resulting equations is

$$\frac{P_0(t+\Delta t) - P_0(t)}{\Delta t} = \mu P_1(t) - \lambda P_0(t) + \frac{0(\Delta t)}{\Delta t}$$

Taking limit on both sides as $\Delta t \to 0$, we get

$$P_0(t) = \frac{d}{dt}\{P_0(t)\} = \mu P_1(t) - \lambda P_0(t);\ n = 0 \qquad ...(2)$$

Step 2: Obtain the System of Steady-State Equations

In the steady-state, P_n (t) is independent of time t, and the rate of change P_n (t) can be considered to be zero. That is,

$$\lim_{t \to \infty} P_n(t) = Pn$$

$$\text{and } \lim_{t \to \infty} \frac{d}{dt}\{P_n(t)\} = 0;\ n = 0, 1, 2, ...$$

Consequently, Eqn. (1) and (2) may be written as:

$$\lambda P_{n-1} + \mu P_{n+1} - (\lambda + \mu) P_n = 0;\ n \geq 1 \qquad (3)$$
$$\mu P_1 - \lambda P_0 = 0;\ n = 0 \qquad (4)$$

Thus these equation constitute the system of steady-state difference equations. The solutions of these equations can be obtained either by (i) iterative method, (ii) generating functions method, or (iii) linear operator method. Here we shall be using the iterative method to find the values of P_1, P_2, ... in terms of P_0, λ and μ.

Step 3: Solve the system of difference equations

From Eqn. (4), we get

$$\lambda P_0 = \mu P_1 \text{ or } P_1 = \left(\frac{\lambda}{\mu}\right) P_0;\ n = 0$$

Putting n = λ in Eqn. (3), we get

$$0 = -(\lambda + \mu)\, P_1 + \lambda\, P_0 + \mu\, P_2$$

or
$$P_2 = \left(\frac{\lambda+\mu}{\mu}\right)P_1 - \frac{\lambda}{\mu}P_0 = \left(\frac{\lambda+\mu}{\mu}\right)\frac{\lambda}{\mu}P_0 - \frac{\lambda}{\mu}P_0 = \left(\frac{\lambda}{\mu}\right)^2 P_0$$

$$P_3 = \left(\frac{\lambda+\mu}{\mu}\right)P_2 - \left(\frac{\lambda}{\mu}\right)P_1$$

$$= \left(\frac{\lambda+\mu}{\mu}\right)\left(\frac{\lambda}{\mu}\right)^2 P_0 - \left(\frac{\lambda}{\mu}\right)\left(\frac{\lambda}{\mu}\right)P_0 = \left(\frac{\lambda}{\mu}\right)^3 P_0;\ n = 2$$

In general, by using the inductive principle, we get

P_n = probability of being in state n (*i.e.*, n customers in the system)

$$= \left(\frac{\lambda}{\mu}\right)^n P_0;\ n = 1, 2, \ldots$$

To obtain the value of P_0, we make use of the fact that (sum of all probabilities)

$$\sum_{n=0}^{\infty} P_n = P_0 + P_1 + \ldots + P_n + \ldots = 1$$

or
$$\sum_{n=0}^{\infty} (\lambda/\mu)^n\ P_0 = 1$$

or
$$P_0 = \frac{1}{\sum_{n=0}^{\infty}(\lambda/\mu)^n};\quad \frac{\lambda}{\mu} < 1$$

The denominator of this expression is an infinite geometric series, whose sum is:

$$\sum_{n=0}^{\infty}(\lambda/\mu)^n = \frac{1}{1-(\lambda/\mu)}$$

and hence $P_0 = 1 - \dfrac{\lambda}{\mu} = 1 - \rho$

and $P_n = \left(\dfrac{\lambda}{\mu}\right)^n\left(1-\dfrac{\lambda}{\mu}\right) = \rho_n\,(1 - \rho);\ \rho < 1,\ n = 0, 1, 2, \ldots$

P_w = Probability than an arriving customer has to wait = $1-P_0 = \lambda/\mu$.

This expression gives the required probability distribution of exactly n customers in the queuing system. Here, P_0 denotes the probability of the system being empty (no customer).

Step 4: Obtain Probability Density Function of Waiting Time Excluding Service Time Distribution

The waiting time distribution of each customer in the steady-state is same, and it is a continuous random variable except that there is a non-zero probability that the delay will be zero, *i.e.,* waiting time is zero. Let w be the time required by the sever to serve all the customers present in the system at a particular time in the steady state. Let ϕ_w (t) be the probability distribution function of w, *i.e.,* $\phi_w (t) = P(w \leq t),\ 0 \leq t \leq \infty$.

If an arriving customer find n (≥ 1) customers already in the system, then in order for a customer to get service at time between 0 and t, all the customers must have been serviced by time t. Let $s_1, s_2, \ldots, s_n$ denote service time of n customers, respectively. Thus

$$w = 0 \;;\; n = 0$$

$$\sum_{i=1}^{n} s_i \;;\; n \geq 1$$

The distribution function of waiting time w for a customer who has to wait is given by

$$P_0 = 1 - \rho \;;\; n = 0,\ t = 0$$

$$P\ \{w \leq t\} = P\left\{\sum_{i=1}^{n} s_i \leq t\right\} \;;\; n \geq 1,\ t > 0.$$

Since the service time for each customer is independent and identically distributed, therefore its probability density function is, $\psi_s (t) = \mu e^{-\mu t}$, $t > 0$, where μ is the mean service rate. Thus,

$$\phi_n (t) = \sum_{i=1}^{n} [P_n \operatorname{Prob}\{(n-1)\text{customers got service at time t}\}]$$

$$\times \text{ Prob } \{\text{one customer is under service during time } \Delta t\}$$

$$= \sum_{i=1}^{n} \left(1 - \frac{\lambda}{\mu}\right)\left(\frac{\lambda}{\mu}\right)^n \left[\frac{(\mu t)^{n-1} e^{-\mu t}}{(n-1)!}\right] \mu\, \Delta t$$

Thus the expression for ϕ_w (t) may be written as

$$\phi_w (t) = P\{w \leq t\} = 1 - \rho;\ t = 0$$

$$\sum_{n=1}^{\infty} P_n \int_0^t \phi_n(t) dt$$

$$= 1 - \rho;\ t = 0$$

$$\sum_{n=1}^{\infty}\rho^n(1-\rho)\int_0^1\frac{(\mu t)^{n-1}.\mu e^{-\mu t}dt}{(n-1)!}; t > 0$$

$= 1 - \rho; t = 0$

$$\int_0^t \rho(1-\rho)^{me^{-\mu t}} dt\sum_{n=1}^{\infty}\frac{(\mu\rho t)^{n-1}}{(n-1)!}; t > 0$$

$= 1 - \rho; t = 0$

$$\rho(1 - \rho)\int_0^1 \mu e^{-\mu(1-\rho)t}dt$$

This shows that the waiting time distribution is discontinuous at t = 0 and continuous in the range $0 < t < \infty$. Thus expression for ϕ_w (t) may also be written as

$$\phi_w (0) = 1 - r; t = 0$$

$$\phi,w (t) = \lambda (1 - \rho) e^{-\mu(1-\rho)t} dt$$

$$= 1\left(1-\frac{\lambda}{\mu}\right)e^{-(\mu-\lambda)t} dt; t > 0$$

Step 5: Calculate Busy Period Distribution

For the busy period distribution, let the ran down variable w denote the total time (waiting and service) that a customer spends in the system. Then the probability density function for the distribution is given by

$$\phi (w : w > 0) = \frac{\phi(w)}{Prob(w > 0)}; \phi(w) = \frac{d}{dt}\{\phi_w(t)\}$$

$$= \frac{\phi(w)}{\int_0^{\infty}\phi(t)dt} = \frac{\lambda\left(1-\frac{\lambda}{\mu}\right)e^{-(\mu-\lambda)t}}{\int_0^{\infty}\lambda\left(1-\frac{\lambda}{\mu}\right)e^{-(\mu-\lambda)t}}dt$$

$$= \frac{\lambda\left(1-\frac{\lambda}{\mu}\right)e^{-(\mu-\lambda)t}}{\frac{\lambda}{\mu}} = (\mu - \lambda) e^{-(\mu-\lambda)t}; t = 0$$

Now $$\int_0^{\infty}\phi(w:w > 0)dt = \int_0^{\infty}(\mu - \lambda)e^{-(\mu-\lambda)t} dt = 1$$

which is required distribution of busy period.

Remark : If l is viewed as a measurement of demand for service and m as the capacity of the service facility, then m – l represents the 'excess capacity' of the system to fill the demand.

Performance Measures for Model I

1. (a) Expected number of customers in the system (customers in the line plus the customer being served)

$$L_s = \sum_{n=0}^{\infty} nP_n = \sum_{n=0}^{\infty} n(1-\rho)\rho^n,\ 0 < \rho < 1$$

$$= (1-\rho)\sum_{n=0}^{\infty} n\rho^n = \rho\,(1-\rho)\sum_{n=0}^{\infty} n\rho^{n-1}$$

$$= \rho\,(1-\rho)\,\{1 + 2\rho + 3\rho^2 + ...\}$$

$$= (1-\rho)\left[\frac{\rho}{(1-\rho)^2}\right] \text{[sum of an arithmetic-geometric series]}$$

or $$L_s = \frac{\rho}{1-\rho} = \frac{\lambda}{\mu-\lambda};\ \rho = \frac{\lambda}{\mu}$$

(b) Expected number of customers waiting in the queue (*i.e.*, queue length)

$$L_q = \sum_{n=1}^{\infty}(n-1)P_n = \sum_{n=1}^{\infty} nP_n - \sum_{n=1}^{\infty} P_n$$

$$= \sum_{n=1}^{\infty} nP_n\left[\sum_{n=0}^{\infty} P_n - P_0\right]$$

$$= L_s - (1 - P_0) = \frac{\lambda}{\lambda-\mu} - \frac{\lambda}{\mu};\ 1 - P_0 = \frac{\lambda}{\mu}$$

or $$L_q = \frac{\lambda^2}{\mu(\mu-\lambda)}$$

2. (a) Expected waiting time for a customer in the queue

$$W_q = \int_0^{\infty} t\cdot\left\{\frac{d}{dt}\phi_W(t)\right\}dt$$

$$= \int_0^{\infty} t\cdot\lambda(1-\rho)e^{-\mu(1-\rho)t}\,dt$$

Integrating by parts $$W_q = 1\,(1-\rho)\left[\frac{te^{-\mu(1-\rho)t}}{-\mu(1-\rho)} - \frac{e^{-\mu(1-\rho)t}}{\mu^2(1-\rho)^2}\right]_0^{\infty}$$

$$= 1\left(1-\frac{\lambda}{\mu}\right)\frac{1}{(\mu-\lambda)^2}$$

or $$W_q = \frac{\lambda}{\mu(\mu-\lambda)} = \frac{L_q}{\lambda}$$

(b) Expected waiting time for a customer in the system (waiting and service)

W_s = Expected waiting time in queue + Expected service time

$$= W_q + \frac{1}{\mu} = \frac{\lambda}{\mu(\mu-\lambda)} + \frac{1}{\mu}$$

or $$= \frac{1}{\mu-\lambda} = \frac{L_s}{\lambda}$$

3. Probability that the number of customers in the system is greater than or equal to k

$$P(n \geq k) = \left(\frac{\lambda}{\mu}\right)^k ; \; P(n > k) = \left(\frac{\lambda}{\mu}\right)^{k+1}$$

4. The variance (fluctuation) f queue length

$$\text{Var (n)} = \sum_{n=1}^{\infty} n^2 P_n - \left(\sum_{n=1}^{\infty} nP_n\right)^2$$

$$= \sum_{n=1}^{\infty} n^2 P_n - (L_s)^2 = \sum_{n=1}^{\infty} n^2 (1-\rho)\rho^n - \left(\frac{\rho}{1-\rho}\right)^2$$

$$= (1 - \rho)\,[1.\rho^2 + 2^2.\rho^2 + 3^2.\rho^3 + ...] - \left(\frac{\rho}{1-\rho}\right)^2$$

or $$\text{Var (n)} = \frac{\rho}{(1-\rho)^2} = \frac{\lambda\mu}{(\mu-\lambda)^2}$$

5. Probability that the queue is non-empty

$$P\,(n > 1) = 1 - P_0 - P_1$$

$$= 1 - \left(1-\frac{\lambda}{\mu}\right) - \left(1-\frac{\lambda}{\mu}\right)\left(\frac{\lambda}{\mu}\right) = \left(\frac{\lambda}{\mu}\right)^2$$

6. Probability of k or more customers in the system

$$P\,(n \geq k) = \sum_{n=k}^{\infty} P_k = \sum_{n=k}^{\infty} (1-\rho)\rho^k$$

$$= (1 - \rho)\,\rho^k \sum_{n=k}^{\infty} \rho^{n-k}$$

$$= (1 - \rho)\,\rho^k\,[1 + \rho + \rho^2 + ...] = \frac{(1-\rho)\rho^k}{(1-\rho)} = \rho^k$$

or $$P\,(n \geq k) = \left(\frac{\lambda}{\mu}\right)^k ; \; P\,(n > k) = \left(\frac{\lambda}{\mu}\right)^{k+1}$$

7. Expected length of non-empty queue

$$L_b = \frac{\text{Expected length of waiting line}}{\text{Prob}(n > 1)}$$

$$= \frac{L_q}{P(n>1)} = \frac{\lambda^2/\mu(\mu-\lambda)}{(\lambda/\mu)^2} = \frac{\mu}{\mu-\lambda}$$

8. Probability of an arrival during the service time when system contains r customers

$$P\ (n = r) = \int_0^{\infty} P_r(t)s(t).dt = \int_0^{\infty} \frac{(\lambda t)^r e^{-\lambda t}}{r!}.\mu e^{-\mu t}\, dt$$

$$= \frac{\lambda^r \mu}{r!}\int_0^{\infty} e^{-(\lambda+\mu)t}.t^r\, dt = \frac{\lambda^r \mu}{r!}\,\frac{\Gamma(r+1)}{(\lambda+\mu)^{r+1}}$$

$$= \left(\frac{\lambda}{\lambda+\mu}\right)^r\left(\frac{\mu}{\lambda+\mu}\right);\ \text{since} \int_0^{\infty} e^{-xt}t^n\, dt$$

$$= \frac{\Gamma(n+1)}{x^{n+1}};\ \Gamma\ (r + 1) = r!$$

Model II {(M/M/l):(∞/SIRO)}

This model is identical to the Model I with a difference only in queue discipline. Since the derivation of P_n is independent of any specific queue discipline, therefore in this model also we have

$$P_n = (1 - \rho)\ \rho^n;\ n = 1, 2, \ldots$$

Consequently, other results will also remain unchanged in any queuing system as long as P_n remains unchanged.

Model III {(M/M/l):(N/FCFS)} Exponential Service, Finite (or limited) Queue

This model is different from Model I in respect of the capacity of the system. Suppose that no more than N customers can be accommodated at any time in the system due to certain reasons. Thus any customer arriving when the system is already contains N customers does not enter the system and is lost. For example, a finite queue may arise due to physical constraint such as emergency room in a hospital; one man barber shop with certain number of chairs for waiting customers, etc.

The difference equations derived in Model I will also be same for this model as long as n < N. The system of steady-state difference equations for this model are

$$\lambda P_0 = \mu P_1 \qquad ; n = 0$$

$$(\lambda + \mu) P_n = \lambda P_{n-1} + \mu P_{n+1} \qquad ; n = 1, 2, \ldots, N-1$$

$$\lambda P_{N-1} = \mu P_N \qquad ; n = N$$

Assumptions of this model are same as that of Model I except that the length of the queue be limited. Moreover, in this case the service rate does not have to exceed arrival rate ($\mu > \lambda$) in order to obtain steady state conditions.

Using the usual procedure, from first two difference equations the probability of a customer in the system for n = 0, 1, 2, ... , N are obtained as follows:

$$P_n = (\lambda/\mu)^n P_0; \; n \leq N$$

Now in order to obtain the value of P_0 use the fact that

$$1 = \sum_{n=0}^{N} P_n = \sum_{n=0}^{N} (\lambda/\mu)^n P_0$$

$$= P_0 \sum_{n=0}^{N} (\lambda/\mu)^n = P_0 \sum_{n=0}^{N} \rho^n$$

$$= P_0 [1 + \rho + \rho^2 + \ldots + \rho^N] = P_0 \left[\frac{1-\rho^{N+1}}{1-\rho}\right]$$

and consequently, $P_0 = \dfrac{1-\rho}{1-\rho^{N+1}}$; $\rho \neq 1$ and $\rho < 1$; $\rho = \lambda/\mu$

$$P_n = \left(\frac{1-\rho}{1-\rho^{N+1}}\right)\rho^n \qquad ; n \leq N;\ \rho \neq 1\ (\lambda \neq \mu)$$

$$\frac{1}{N+1} \qquad ; \rho = 1\ (\lambda = \mu)$$

The steady-state solution in this case exists even for $\rho > 1$. This is due to the limited capacity of the system which controls the arrivals by the queue length (= N – 1) not by the relative rates of arrival and departure, λ and μ. If $\lambda < \mu$ and $N \to \infty$, then $P_n = (1 - \lambda/\mu)(\lambda/\mu)^n$, which is same is in Model I.

Performance Measures for Model III

1. Expected number of customers in the system

$$L_s = \sum_{n=1}^{N} nP_n = \sum_{n=1}^{N} n\left(\frac{1-\rho}{1-\rho^{N+1}}\right)\rho^n$$

$$= \frac{1-\rho}{1-\rho^{N+1}} \sum_{n=0}^{N} n\rho^n = \frac{1-\rho}{1-\rho^{N+1}}(\rho + 2\rho^2 + 3\rho^3 + \ldots + N\rho^N)$$

$$= \frac{\rho}{1-\rho} - \frac{(N+1)\rho^{N+1}}{1-\rho^{N+1}}; \; \rho \neq 1 \; (\lambda \neq \mu)$$

$$\frac{N}{2}; \; \rho = 1 \; (\lambda = \mu)$$

2. Expected queue length or expected number of customers waiting in the system

$$L_q = L_s - \frac{\lambda}{\mu} = L_s - \frac{\lambda(1-P_N)}{\mu}$$

3. Expected waiting time of a customer in the system (waiting + service)

$$W_s = \frac{L_s}{\lambda(1-P_N)}$$

4. Expected waiting time of a customer in the queue

$$W_q = W_s - \frac{1}{\mu} \text{ or } \frac{L_q}{\lambda(1-P_N)}$$

5. Fraction of potential customers lost (= fraction of time system is full)

$$P_N = P_0 \, \rho^N$$

Effective arrival rate, $\lambda_e = \lambda \; (1 - P_N)$

Effective traffic intensity, $\rho_e = \lambda_e/\mu$.

PROBABILITY DISTRIBUTIONS IN QUEUING SYSTEMS

It is assumed that customers joining a queuing system arrive in random manner and follow a Poisson distribution or equivalently the inter-arrival times follow exponential distribution.

In most of the cases, service times are also assumed to be exponentially distributed. It implies that the probability of service completion in any short-time period is constant and independent of the length of time that the service has been in progress. The basic reason for assuming exponential service is that it helps in formulating simple mathematical models which ultimately help in analysing a number of aspects of queuing problems.

The number of arrivals and departures (those served) during an interval of time in a queuing system is controlled by the following assumptions (also called *axioms*).

(i) The probability of an event (arrival or departure) occurring during the time interval $(t, t + \Delta t)$ depends on the length of time interval Δt. That is, probability of the event does not depend either on number of events that occur upto time t or the specific value of t, meaning that the events that occur in non-overlapping time are statistically independent.

(ii) The probability of more than one event occuring during the time interval $(t, t + \Delta t)$ is negligible.

It is denoted by $0(\Delta t)$.

(iii) Almost one event (arrival or departure) can occur during a small time interval Δt. The probability of an arrival during the time interval $(t, t + \Delta t)$ is given by

$$P_1(\Delta t) = \lambda \Delta t + 0(\Delta t).$$

where λ is a constant and independent of the total number of arrivals up to time t; λt is a small time interval and $0(\Delta t)$ represents the quantity that becomes negligible when compared to Δt as $\Delta t \rightarrow 0$, *i.e.*,

$$\lim_{\Delta t \to 0} \{0(\Delta t)/\Delta t\} = 0.$$

DISTRIBUTION OF ARRIVALS (PURE BIRTH PROCESS)

The arrival process assumes that the customers arrive at the queuing system and never leave it. Such a process is called *Pure Birth Process*. The aim is to derive an expression for the probability $P_n(t)$ of n arrivals during time interval $(t, t + \Delta t)$. The terms commonly used in the development of various queuing models are the following:

Δt = a time interval so small that the probability of more than one customer's arrival is negligible, *i.e.*, during any given small interval of time Δt only one customer can arrive.

$\lambda \Delta t$ = probability that a customer will arrive in the system during time Δt.

$1 - \lambda \Delta t$ = probability that no customer will arrive in the system during time $\lambda \Delta t$.

Remark : If the arrivals are completely random, then the probability distribution of a number of arrivals in a fixed time interval follows a Poisson distribution.

Distribution of Inter-Arrival Times (Exponential Process)

If the number of arrivals, n, in time t follows the Poisson distribution, then

$$P_n(t) = \frac{(\lambda t)^n}{n!} e^{-\lambda t}, \quad n = 0,1,2,....$$

is an associated random variable defined as the inter-arrival time T follows the exponential distribution $f(t) = \lambda e^{-\lambda t}$ and *vice-versa*.

Markovian Property of Inter-Arrival Times : The markovian property of inter-arrival times states that the probability that a customer currently in service is completed at some time t is independent of how long he has already been in service. That is

$$\text{Prob } \{T \geq t \mid T \geq t_0\} = \text{Prob } \{0 \leq t \mid T \leq t_0\}$$

where T is the time between successive arrivals.

Distribution of Departures (Pure Death Process)

The departure process assumes that no customer joins the system while service is continued for those who are already in the system. Let, at time t = 0 (starting time) there be $N \geq 1$ customers in the system. Since service is being provided at the rate of μ, therefore, customers leave the system at the rate u after being serviced. Such a process is called *pure death process*.

Basic Axioms

(i) Probability of the departure during time Δt is $\mu \Delta t$.

(ii) Probability of more than one departure between time t and $t + \Delta t$ is negligible.

(iii) The number of departures in non-overlapping intervals are statistically independent.

The following terms are used in the development of various queuing models.

$\mu \Delta t$ = Probability that a customer in service at time t will complete service during time Δt.

$1 - \mu At$ = Probability that the customer in service at time t will not complete service during time Δt.

Distribution of Service Times

The probability density function s(f) of service time is given by:

$$s(t) = \begin{cases} \mu e^{-\mu t} & ; \ 0 \leq t \leq \infty \\ 0 & ; \ t , 0 \end{cases}$$

This shows that service times follows negative exponential distribution with mean $1/\mu^2$ and variance l/μ^2.

From Fig. 16.3(a) it can be seen that service times of short duration have the highest probability of occurrence. As the service time increases, the probability of occurrence tails off (exponentially) towards zero probability. The area under the negative exponential distribution curve is determined as:

$$F(T) = \int_0^t \mu e^{-\mu t}\, dt = \left[-\mu e^{-\mu t}\right]_0^t$$

$$= e^{-\mu t} + e^0 = 1 - e^{-\mu t}$$

It is also described as:

$$F(T) = f(t \geq T) = e^{-\mu t}$$

where F(T) is the area under the curve to the left of T. Thus,

$$1 - F(T) = f(t \geq T) = e^{-\mu t}$$

is the area under the curve to the right of T.

For example, if mean service time ($1/\mu$.) at a service station is 2 minutes, then probability that service will take Tor more minutes is shown below:

Service times of at least T :	0	1	2	3	4	5
Probability $f(t \geq T)$:	1	0.607	0.368	0.223	0.135	0.082

TRANSIENT AND STEADY STATES OF THE SYSTEM

Queuing theory analysis involves the study of system's behaviour over time. If the operating characteristics (behaviour of the system) vary with time, it is said to be in *Transient State*. Usually a system is transient during the early stages of its operation, when its behaviour still depends upon the initial conditions. However, it is the "long-run" behaviour or the steady state condition of the system which is more important. A system is said to be in steady state condition if its behaviour becomes independent of time.

An essential condition for reaching a steady stale is that the total elapsed time since the start of the operation must be sufficiently large (theoretically, it should tend to infinity). However, this is not the sufficient condition as the parameters of the system also affect its said.

For example, if the average arrival rate is less than average service rate and with are constant, the system eventually settles down to a steady state and the probability of finding a particular length of queue will be sea at any time. If the rates are not constant, the system will not reach a steady state, but it could remain stable. If the arrival rate is greater than service rate, the system cannot attain a steady state (regardless of the length of elapsed time); it is rather unstable, queue length increases steadily with time and theoretically, it could build up to infinity.

Such state of the system is called *explosive state*. Evidently, imposing a limit on the maximum length of the queue (so that further arrivals are not accepted) automatically ensures stability. Queuing situations which are unstable for a limited time are common in practice—rush hour traffic is an example.

In this text we shall consider the steady state analysis; transient and explosive states require complex mathematical tools for analysis and will not be touched upon.

MULTI-SERVER QUEUING MODELS

Model IV {(M/Ms):(∞/FCFS)} Exponential Service—Unlimited Queue

This model is an extension of Model I. In this case instead of single service channel, there are multiple servers in parallel equal to s. For this queuing system it is assumed that customers arrive according to a Poisson process at an average rate of λ customers per unit of time and are served on a first-come, first-served basis at any of the servers. These servers are identical, each serving customers according to an exponential distribution with an average of μ customers per unit of time. It is further assumed that only one queue is formed.

If n < s, (number of customers in the system is less than the number of servers), then there will be no queue. However, (s – n) number of servers are not busy. The combined service rate will then be:

$$m_n = nm;\ n < s$$

Step 1: Obtain the System of Differential-Difference Equations

By the same arguments as is Model I, we have

$$P_n(t + \Delta t) = P_n(t)\{1 - \lambda\,\Delta t\}\{1 - n\,\lambda\,\Delta t\} + P_{n+1}(t)\ \{1 - \lambda\,\Delta t\}\{(n + 1)\ \mu\,\Delta t\} + P_{n-1}\,(t)\ \{\lambda\,\Delta t\}\ \{1 - (n - 1)\ \mu\,\Delta t\}$$

$$= -\ (\lambda + n\mu)\ P_n\ (t)\ \Delta t + (n + 1)\ \lambda\ P_{n+1}\ (t)\ \Delta t + \lambda\ P_{n-1}\ (t)\ \Delta t + P_n\ (t) + \text{terms involving } (\Delta t)^2;\ 1 \geq n < s$$

$$P_n\ (t + \Delta) = P_n\ (t)\ \{1 - \lambda\,\Delta t\}\ \{1 - \lambda\,\Delta t\} + P_{n+1}\ (t)\ \{1 - \lambda\,\Delta t\}\ \{s\mu\,\Delta t\} + P_{n-1}\ (t)\ \lambda\,\Delta t\ \{1 - s\ \mu\,\Delta t\}$$

$$= -\ (\lambda + s\mu)\ P_n\ (t)\ \Delta t + s\mu\ P_{n+1}\ (t)\ \Delta t + \lambda\ P_{n-1}\ (t)\ \Delta t + P_n\ (t) + \text{terms involving } (\Delta t)^2;\ n \geq s$$

and $P_0\ (t + \Delta t) = P_0\ (t)\ (1 - \lambda\,\Delta t) + P_1\ (t)\ \lambda\,\Delta t;\ n = 0$

Dividing these equations by Δt and then taking limit as $\Delta t \to 0$, we get

$$P'_n\ (t) = -\ (\lambda + n\mu)\ P_n\ (t) + (n + 1)\ \mu\ P_{n+1}\ (t) + \lambda\ P_{n-1}\ (t);\ 1 \leq n < s$$

$$P'_n\ (t) = -\ (\lambda + s\mu)\ P_n\ (t) + s\mu\ P_{n+1}\ (t) + \lambda\ P_{n-1}\ (t);\ n \geq s$$

and $P'_0\ (t) = -\ l\ P_0\ (t) + \mu\ P_1\ (t);\ n = 0.$

Step 2: Obtain the System of Steady-State Equations

In the steady-state condition, the differential-difference equations are obtained from the above equations at $t \to \infty$. This yields

$$- \lambda P_0 + \mu P_1 = 0;\ n = 0$$
$$- (\lambda + n\mu) P_n + (n + 1) \mu P_{n+1} + \lambda P_{n-1} = 0;\ 1 \le n < s$$
$$- (\lambda + s\mu) P_n + s\mu P_{n+1} + \lambda P_{n-1} = 0;\ n \ge s$$

Step 3: Solve the System of Difference Equations

By applying the iterative method as discussed earlier, the probability of n customers in the system is given by

$$P_n = \frac{\rho^n}{n!} P_0;\ 1 \le n < s$$
$$\frac{\rho^n}{s!s^{n-s}} P_0;\ n \ge s;\ \rho = \lambda/\mu$$

In order to find P_0, we use the condition

$$1 = \sum_{n=0}^{\infty} P_n = \sum_{n=0}^{s-1} P_n + \sum_{n=s}^{\infty} P_n = \sum_{n=0}^{s-1} P_n \frac{1}{n!}\left(\frac{\lambda}{\mu}\right)^n P_0 + \sum_{n=s}^{\infty} \frac{1}{s!s^{n-s}}\left(\frac{\lambda}{s\mu}\right)^n P_0$$

$$= P_0 \left[\sum_{n=0}^{s-1} \frac{s^n}{n!}\left(\frac{\lambda}{s\mu}\right)^n + \sum_{n=s}^{\infty} \frac{s^n}{s!s^{n-s}}\left(\frac{\lambda}{s\mu}\right)^n P_0\right]$$

$$= P_0 \left[\sum_{n=0}^{s-1} \frac{(s\rho)^n}{n!} + \frac{s^s}{s!}\sum_{n=s}^{\infty} \rho^n\right] = P_0 \left[\sum_{n=0}^{s-1} \frac{(s\rho)^n}{n!} + \frac{s^s}{s!}\frac{\rho^s}{1-\rho}\right];\ \rho = \frac{\lambda}{s\mu}$$

[Since $\sum_{n=s}^{\infty} \rho^n = \rho^{s} + \rho^{s+1} + \ldots = \rho^{s}/(1-\rho)$, sum of infinite G.P.; $\rho < 1$]

Thus the probability that the system shall be idle is

$$P_0 = \left[\sum_{n=0}^{s-1} \frac{(s\rho)^n}{n!} + \frac{1(s\rho)^s}{s!1-\rho}\right]^{-1};\ \rho = \lambda/s\mu$$

$$= \left[\sum_{n=0}^{s-1} \frac{1}{n!}\left(\frac{\lambda}{\mu}\right)^n + \frac{1}{s!}\left(\frac{\lambda}{\mu}\right)^n \frac{s\mu}{s\mu - \lambda}\right]^{-1}$$

Performance Measures for Model IV

1. Expected number of customers waiting in the queue (length of line)

$$L_q = \sum_{n=s}^{\infty}(n-s)P_n = \sum_{n=s}^{\infty}(n-s)\frac{\rho^n}{s^{n-s}.s!}P_0$$

$$= \frac{\rho^s P_0}{s!}\sum_{n=s}^{\infty}(n-s)\rho^{n-s} \ ; \ \rho = \frac{\lambda}{\mu}$$

$$= \frac{\rho^s P_0}{s!}\sum_{m=0}^{\infty}m\rho^m \ ; \ n - s = m$$

$$= \frac{\rho^s}{s!}\cdot\rho P_0\sum_{m=0}^{\infty}m\rho^{m-1} = \frac{\rho^s}{s!}\cdot\rho P_0\frac{d}{d\rho}\left[\sum_{m=1}^{\infty}\rho^m\right]$$

$$= \frac{\rho^s}{s!}\rho P_0\frac{1}{(1-\rho)^2} = \left[\frac{1}{s!}\left(\frac{\lambda}{\mu}\right)^s\frac{\lambda.s\mu}{(s\mu-\lambda)^2}\right]P_0 \ .$$

or $$L_q = \left[\frac{1}{(s-1)!}\left(\frac{\lambda}{\mu}\right)^s\frac{\lambda\mu}{(s\mu-\lambda)^2}\right]P_0$$

2. Expected number of customers in the system

$$L_s = L_q + \frac{\lambda}{\mu}$$

3. Expected waiting time of a customer in the queue

$$W_q = \frac{L_q}{\lambda} = \left[\frac{1}{(s-1)!}\left(\frac{\lambda}{\mu}\right)^s\frac{\mu}{(s\mu-\lambda)^2}\right]P_0$$

4. Expected waiting time that a customer spends in the system

$$W_s = W_q + \frac{1}{\mu} = \frac{L_q}{\lambda}+\frac{1}{\mu}$$

5. Probability that an arriving customer has to wait (busy period)

$$P\ (n \geq s) = \sum_{n=s}^{\infty}P_n = \sum_{n=s}^{\infty}\frac{1}{s!s^{n-s}}\left(\frac{\lambda}{\mu}\right)^n P_0$$

$$= \frac{1}{s!}\left(\frac{\lambda}{\nu}\right)^s P_0\sum_{m=0}^{\infty}\left(\frac{\lambda}{\mu}\right)$$

or $$P\ (n \geq s) = \frac{1}{s!}\left(\frac{\lambda}{\mu}\right)^s P_0\left[\frac{1}{1-(\lambda/\mu)}\right], \text{ (sum of infinite G.P).}$$

Model V {M/M/s) : (N/FCFS)} Exponential Service–Limited (Finite) Queue

This model is the further extension of Model IV, the assumption of an unlimited waiting area for customers is not valid in certain cases such as:

(i) The parking area once full to its capacity, turns away arriving vehicles. Here arriving vehicles are customers, each parking space is a server and there is no waiting space

(ii) In a production facility, parts arriving from a previous production stage to a machine for further the production at the previous stage must come to halt. He arriving parts are the customers, and the machine is the server. Once the waiting area is full arriving customers are turned away and may or may not return. Thus, in this case while performing economic analysis, cost associated with losing a customer is also taken into consideration along with the cost per server and the cost of waiting. Let.

$$\lambda_n = \begin{cases} \lambda & ; \quad n \le N \\ 0 & ; \quad n > N \end{cases}$$

and $\mu_n = \begin{cases} n\mu & ; \quad n < N \\ s\mu & ; \quad s \le n \le N \end{cases}$

Probability of n customers in the system in the steady-state condition is

$$P_n = \begin{cases} \dfrac{1}{n!}\left(\dfrac{\lambda}{\mu}\right)^n P_0 & ; n \le s \\ \dfrac{1}{s!s^{n-s}}\left(\dfrac{\lambda}{\mu}\right)^n P_0 & ; s < n \le N \\ 0 & ; n > N \end{cases}$$

The condition $\displaystyle\sum_{n=0}^{N} P_n = \sum_{n=0}^{s-1} P_n + \sum_{n=s}^{N} P_n = 1$

gives $$P_0 = \left[\sum_{n=0}^{s-1} \frac{1}{n!}\left(\frac{\lambda}{\mu}\right)^n + \sum_{n=s}^{N} \frac{1}{s!s^{n-s}}\left(\frac{\lambda}{\mu}\right)^n\right]^{-1}$$

or $$\sum_{n=0}^{s-1} \frac{1}{n!}\left(\frac{\lambda}{\mu}\right)^n + \sum_{n=s}^{N} \frac{1}{s!s^{n-s}}\left(\frac{\lambda}{\mu}\right)^n P_0 = 1$$

$$P_0\left[\sum_{n=0}^{s-1}\frac{s^n}{n!}\left(\frac{\lambda}{s\mu}\right)^n+\sum_{n=s}^{N}\frac{s^n}{s!s^{n-s}}\left(\frac{\lambda}{s\mu}\right)^n\right]=1$$

$$P_0\left[\sum_{n=0}^{s-1}\frac{(s\rho)^n}{n!}+\frac{(s\rho)^s}{s!}\sum_{n=0}^{N-s}\rho^n\right]=1\ ;\ \rho=\lambda/s\,\mu$$

Here $\displaystyle\sum_{n=0}^{N-s}\rho^n=\frac{1-\rho^{N-s+1}}{1-\rho}$; $\rho\ (=\lambda/s\mu)\neq 1$,

(sum of GP of N – s + 1 terms)

$N-s+1\ ;\ \rho=1$

Thus the probability, P_0 that the system shall be idle is

$$P_0=\left[\sum_{n=0}^{s-1}\frac{(s\rho)^n}{n!}+\frac{1}{s!}\left(\frac{\lambda}{\mu}\right)^s\frac{s\mu}{s\mu-\lambda}\left\{1-\left(\frac{\lambda}{s\mu}\right)^{N-s+1}\right\}\right]^{-1}\ ;\ \rho=\lambda/s\mu\ (=1)$$

$$=\left[\sum_{n=0}^{s-1}\frac{(s\rho)^n}{n!}+\frac{1}{s!}\left(\frac{\lambda}{\mu}\right)^s(N-s+1)\right]^{-1}$$

Remarks : 1. For $N\to\infty$ and $\lambda/s\mu<1$, the above result corresponds to that of Model IV.

2. For s = 1, the above results reduces to the form given in Model III.

Performance Measures of Model V

1. Effective arrival rate, $\lambda_e=\lambda\,(1-PN)$

   Effective traffic intensity, $\rho_e=\lambda_e/\mu$

2. Expected number of customers in the queue

$$L_q=\sum_{n=s}^{N}(n-s)P_n=\sum_{n=s}^{N}(n-s)\frac{s^n}{s!s^{n-s}}\left(\frac{\lambda}{s\mu}\right)^n P_0$$

$$=\sum_{n=s}^{N}(n-s)\frac{(s\rho)^n}{s!s^{n-s}}P_0$$

$$=\frac{(s\rho)^s P_0\rho}{s!}\sum_{x=0}^{n-s}x.\rho^{x-1}\ ;\ x=n-s;\ \rho\ /\ \lambda/s\mu.$$

$$=\frac{(s\rho)^s\rho P_0}{s!}\sum_{x=0}^{N-s}\frac{d}{d\rho}(\rho^x)=\frac{(s\rho)^s\rho}{s!}\frac{d}{d\rho}\left[\sum_{x=0}^{N-s}\rho^x\right]P_0$$

$$= \frac{(s\rho)^s \rho}{s!} \frac{d}{d\rho}[1+\rho+\rho^2+...+\rho^{N-s}]P_0$$

$$= \frac{(s\rho)^s \rho}{s!} \frac{d}{d\rho}\left[\frac{1-\rho^{N-s+1}}{1-\rho}\right]P_0, \qquad \text{(GP of N – s + 1 terms)}$$

or
$$L_q = \frac{(s\rho)^s \rho}{s!(1-\rho)^2}[1-\rho^{N-s+1}-(1-\rho)(N-s+1)\rho^{N-s}]P_0$$

3. Expected number of customers in the system

$$L_s = L_q + \left(\frac{\lambda}{\mu}\right)(1-P_N)$$

$$= L_q + s - P_0 \sum_{n=0}^{s-1} \frac{(s-n)}{n!}\left(\frac{\lambda}{\mu}\right)^n$$

4. Expected waiting time in the system

$$W_s = \frac{L_s}{\lambda(1-P_N)}$$

5. Expected waiting time in the queue

$$W_q = W_s - \frac{1}{\mu} = \frac{L_q}{\lambda(1-P_N)}$$

6. Fraction server idle time

$$1 - \frac{L_s - L_q}{s} = 1 - \frac{\rho_e}{s}.$$

Remark : If no queue is allowed, then number of customers intend to join a queuing system should not exceed the number of servers, *i.e.*, $n \leq s$. Thus, we have

$$P_n = \frac{1}{n!}\left(\frac{\lambda}{\mu}\right)^2 P_0$$

and
$$P_0 = \left[\sum_{n=0}^{s} \frac{1}{n!}\left(\frac{\lambda}{\mu}\right)^n\right]^{-1}$$

Since no queue forms, $L_q = W_q = 0$.

The performance measures for this special case of Model V are

1. Fraction of potential customers lost, $P_s = \frac{1}{s!}\left(\frac{\lambda}{\mu}\right)^s P_0$

2. Effective arrival rate, $\lambda_e = \lambda\ (1 - P_s)$
3. Expected waiting time in the system, $W_s = 1/\mu$
4. Expected number of customers in the system, $L_s = \lambda\ W_s = \lambda/\mu$
5. Fraction idle time for server $= 1 - \rho_e/s$.

FINITE CALLING POPULATION QUEUING MODELS

Model VI {(M/M/l) : (M/GD)} Single Server—Finite Population (Source) of Arrivals

This model is similar to the Model I except that the calling population (input source) of potential customers is limited, say M. Thus arrival of additional customers is not allowed to join the system when the system becomes busy in serving the existing customers in the queue. Few applications of this model are as under:

(i) A fleet of office cars available for 5 senior executives. Here these 5 executives are the customers, and the cars in the fleet are the servers.

(ii) A maintenance staff provides repair to M machines in a workshop. Here the M machines are customers and the repair staff members are the servers.

When there are n customers in the system, then system if left with the capacity to accommodate M – n more customers. Thus, further arrival rate of customers to the system will be λ (M = n). That is, for s = 1, the arrival rate and service rate is stated as follows:

$$\lambda_n = \lambda\ (M - n)\ ;\ n = 1, 2, \ldots, M$$
$$0\ ;\ n > N$$
$$\mu_n = \mu\ ;\ n = 1, 2, \ldots, M$$

Performance Measures of Model VI

Substituting for λ_n and μ_n in the expression for P_n and P_0 in Model IV(B), we get

1. Probability that the system is idle

$$P_0 = \left[\sum_{n=0}^{M} \frac{M!}{(M-n)!}\left(\frac{\lambda}{\mu}\right)^n\right]^{-1}$$

2. Probability that there are n customers in the system

$$P_n = \frac{M!}{(M-n)!}\left(\frac{\lambda}{\mu}\right)^n P_0\ ;\ n = 1, 2, \ldots, M$$

3. Expected number of customers in the queue (or queue length)

$$L_q = \sum_{n=1}^{M} (n-1)P_n = M - \left(\frac{\lambda+\mu}{\lambda}\right)(1-P_0)$$

4. Expected number of customers in the system

$$L_s = \sum_{n=0}^{M} nP_n = L_q + (1 - P_0)$$

$$= M - \frac{\mu}{\lambda}(1-P_0)$$

5. Expected waiting time of a customer in the queue

$$W_q = \frac{L_q}{\lambda(M-L_s)}$$

6. Expected waiting time of a customer in the system

$$W_s = W_q + \frac{1}{\mu} \text{ or } \frac{L_s}{\lambda(M-L_s)}$$

Model VII {(M/M/s) : (M/GD)} Multiserver—Finite Population (Source) of Arrivals

When number of servers are more than one (*i.e.,* $s > 1$), the steady-state equations are derived in the same way as in Models IV and V.

$$M\rho\, P_0 = P_1 \;;\; n = 0$$

$$\{(M - n)\,\rho + n\}\, P_n = (M - n + 1)\,\rho\, P_{n-1} + (n + 1)\, P_{n+1} \;;\; 0 \le n \le M$$

$$\{(M - n)\,\rho + s\}\, P_n = (M - n + 1)\,\rho\, P_{n-1} + s\, P_{n+1} \;;\; s < n \le M - 1$$

$$s\, P_M = \rho\, P_{n-1}.$$

We define λ_n and μ_n as the rate of arrival and service, respectively as follows:

$$\lambda_n = (M - n)\,\lambda \;;\; 0 \le n < M,\quad 0 \;;\; n \ge M$$

$$n\mu \;;\; 0 \le n < s$$

$$\mu_n = s\mu \;;\; n \ge s.$$

Again substituting for λ_n and μ_n in the expression for P_0 and P_n is Model V(A), we get

$$P_n = \frac{M!}{n!(M-n)!}\left(\frac{\lambda}{\mu}\right)^n P_0 \;;\; 0 \le n \le s$$

$$\frac{M!}{(M-n)!s!s^{n-s}}\left(\frac{\lambda}{\mu}\right)^n P_0 \;;\; s < n \le M$$

$$P_0 = \left\{\sum_{n=0}^{s-1} \frac{M!}{n!(M-n)}\left(\frac{\lambda}{\mu}\right)^n)\sum_{n=s}^{M} \frac{M!}{n!(M-n)!s!s^{n-s}}\left(\frac{\lambda}{\mu}\right)^n\right\}^{-1}$$

Performance Measures of Model VII

1. Expected number of customers in the queue

$$L_q = \sum_{n=s+1}^{M}(n-s)P_n = \sum_{n=s+1}^{M} nP_n - s\sum_{n=s+1}^{M} P_n$$

$$= \sum_{n=0}^{M} nP_n - \sum_{n=0}^{s} nP_n - s\left\{\sum_{n=0}^{M} P_n - \sum_{n=0}^{s} P_n\right\}$$

$$= \sum_{n=0}^{M} nP_n - \sum_{n=0}^{s} nP_n - s\left\{1 - \sum_{n=0}^{s} P_n\right\}$$

$$= L_s - \bar{s} + \sum_{n=0}^{s}(s-n)P_n = L_s - (s - \bar{s})$$

where $\bar{s}$ = expected number of idle servers = $\sum_{n=0}^{s}(s-n)P_n$

2. Expected number of customers in the system

$$L_s = L_q + (s - \bar{s}) = L_q + \frac{\lambda_{eff}}{\mu}$$

where $\lambda_{eff} = \mu\,(s - \bar{s})$.

The expression for λ_{eff} represents the expected arrival rate $\lambda\,(M - n)$ of customers (n customers are already in the system each with mean arrival rate λ) under steady-state conditions. Thus

$$\lambda_{eff} = \sum_{n=0}^{M}\lambda(M-n)P_n = \lambda\, M \sum_{n=0}^{M} nP_0 - \lambda\sum_{n=0}^{M} nP_n$$

$$= \lambda M - \lambda L_s = \lambda\,(M - L_s)$$

3. Expected waiting time of a customer in the system

$$W_s = \frac{L_s}{\lambda_{eff}} = \frac{L_s}{\lambda(M-L_s)}$$

4. Expected waiting time of customer in the system

$$W_q = \frac{L_q}{\lambda_{eff}} = \frac{L_q}{\lambda(M-L_s)}$$

PERFORMANCE MEASURES OF MODEL VIII

1. Expected number of phases (not customers) in the system

$$L_q(k) = \frac{L_s(k)}{\mu} = \frac{k+1}{2}\frac{\lambda}{\mu(\mu-\lambda)}$$

$$L_s(k) = \frac{k+1}{2}\frac{\lambda}{\mu-\lambda}$$

2. Expected number of customers (not phases) in the queue

$$L_q = \frac{L_s(k) - \text{Average number of phase sin service}}{k}$$

$$= \frac{1}{k}\left[\frac{k+1}{2}\frac{\lambda}{\mu-\lambda} - \frac{k+1}{2}\frac{\lambda}{\mu}\right]$$

$$= \left(\frac{k+1}{2k}\right)\left(\frac{\lambda^2}{\mu(\mu-\lambda)}\right)$$

Since $1/\mu$ is the average service time per customer and $(k + 1)/2$ is the average number of phases of one customer in service, therefore time taken for serving a customer will be $(k + 1)/2\mu$. Thus, average number of phases that arrive in this time would be $\lambda(k + 1)/2\mu$

3. Expected waiting time of a customer in the queue

$$W_q = \frac{L_q}{\lambda} = \frac{k+1}{2k}\frac{\lambda}{\mu(\mu-\lambda)}$$

4. Expected waiting time of a customer in the system

$$W_s = W_q + \frac{1}{\mu}$$

$$= \frac{k+1}{2k}\frac{\lambda}{\mu(\mu-\lambda)} + \frac{1}{\mu}$$

5. Expected number of customers in the system

$$L_s = L_q + \frac{\lambda}{\mu} \text{ or } L_s = \lambda W_s$$

$$= \frac{k+1}{2k}\frac{\lambda^2}{\mu(\mu-\lambda)} + \frac{\lambda}{\mu}$$

SPECIAL PURPOSE QUEUING MODELS

Model IX Single Server, Non-Exponential Service Times Distribution–Unlimited Queue

When service time no longer follows an exponential distribution, the normal distribution could also be used to represent the service pattern of a single server queuing system. A queuing model where arrivals form a Poisson process, while the service times follow normal distribution depends on the standard deviation for service time and assumes no particular form for the distribution itself. The performance measures in this case are determined as under:

$$P_0 = 1 - \frac{\lambda}{\mu}$$

$$L_q = \frac{\lambda^2\sigma^2 + (\lambda\mu)^2}{2(1-\lambda/\mu)}\ ;\ L_s = L_q + \frac{\lambda}{\mu}$$

$$W_q = \frac{L_q}{\lambda}\ ;\ W_s = W_q + \frac{1}{\mu}$$

The values of these performance measures except P_0 depend on standard deviation of the service time distribution. Since σ appears in numerator, therefore, greater variability in the service time will result in longer waiting time. Thus, consistency in service time is very important to the overall quality of the service period.

Model X Single Server, Constant Service Times-Unlimited Queue

If service time is constant ($= \lambda/\mu$) instead of exponential distribution time to serve each customer, then the variance $\sigma^2 = 0$ and obviously, values of L_s, L_q, W_s and W_q will be less than those values in the models discussed before.

Substituting $\sigma^2 = 0$ in Model IX, we get

$$L_q = \frac{(\lambda/\mu)^2}{2\{1-(\lambda/\mu)\}} = \frac{\lambda^2}{2\mu(\mu-\lambda)}\ ;\ L_s = L_q = (\lambda/\mu)$$

$$W_q = \frac{L_q}{\lambda} = \frac{\lambda}{2\mu(\mu-\lambda)}\ ;\ W_s = W_q + \frac{1}{\mu}$$

The assumption of constant service time puts an absolute lower bound on the value of the mean queue length and upper bound can be obtained by assuming exponential service because negative exponential distribution is the high-variance distribution.

Model XI Service Level Cost Model

The service level is the function of two conflicting costs:

(i) Cost of offering the service to the customers

(b) Cost of delay in offering service to the customers

This means the customer's service level determines the efficiency of the queuing system. Hence, an optimum service level that must be maintain by a queuing system can be determined by determining the optimum level of service rate (μ) to offer service and number of service facilities (s, servers) to avoid excessive delay in offering the service.

In this model, we will discuss the procedure of determining the optimum values of both service rate (μ) and number of service.

Optimum Service Rate : The procedure discussed here for determining optimum service rate (μ) is based on Model I. Let

C_1 = cost per server per unit of time, *i.e.*, cost per unit of time having a server available

C_3 = cost of waiting per customer per unit time, *i.e.*, cost per unit of time having a customer wait in the system

TC(μ) = total expected cost of waiting and service per unit time, given the service rate μ.

Then, the cost equation can be written as

TC(μ) = (Service rate × Cost per service per unit time) + (Number of customers in the system × Waiting cost per customer per unit time)

$$= \mu C_1 + L_s C_2$$

$$= \mu C_1 + \frac{\lambda}{\mu - \lambda} C_2; \; L_s = \frac{\lambda}{\mu - \lambda} \text{ (Model I).}$$

Since service rate is measured in continuous units of time, its optimum value can be obtained using the concept of maximum and minima in differential calculus. Thus, we get

$$\frac{d}{d\mu} TC(\mu) = C_1 - \frac{\mu C_2}{(\mu - \lambda)^2} = 0$$

as the necessary condition for maximum or minimum value of μ. Thus, we get

$$\mu = \lambda + \sqrt{\lambda C_2 / C_1}$$

This value of m represents it optimum value because second derivative of TC(μ) with respect to μ is positive for μ > λ.

Remark : If a queuing system has limited capacity of N customers (see Model III), then TC(m) can be modified as:

$$TC(N, \mu) = \mu C_1 + L_s.C_2 + N.C_3 + \lambda P_N.C_4$$

where C3 = cost of serving each additional customer per unit time

C4 = cost per lost (balking) customer (*i.e.*, λP^N), *i.e.*, cost of losing a customer.

Optimum Number of Service Facilities (Servers) : For determining the optimum number of service facilities (servers) we will consider the operational characteristics of Model IV(A): {(M/M/s) : (∞/FCFS)}. In this case cost equation can be written as:

$$TC(s) = sC_1 + L_s(s).\ C_2$$

where C_s = cost per server per unit of time,

C_w = cost of waiting per customer per unit of time,

$L_s(s)$ = expected number of customers with system when there are number of servers.

Since s can be measured only in discrete units, therefore, its optimum value can be obtained by using the concept from difference equations. That is, the necessary condition for a minimum of the given function is

$$\Delta TC(s-1) < 0 < \Delta TC(s) .$$

The condition $\Delta TC(s-1) < 0$, gives

$$TC(s) - TC(s-1) < 0,$$

or $$s.C_s + L_s\ (s).\ C_w - \{(s-1)\ C_s + L_s\ (s-1).\ C_w\}$$

$$L_s\ (s-1) - L_s\ (s) > \frac{C_S}{C_W}$$

The condition AT'C(s) > 0 gives

$$TC(s - l) - TC(s) > 0$$

$$\{(s + l)C, + L_s(s + l)C_s\} - sC_s + L_s(s)C_w\} > 0.$$

These two results together yield the condition to determine optimum value of s.

$$L_s - L_s(s + l) \leq \frac{C_S}{C_W} \leq L_s(s - l) - L_s(s).$$

Input Source Characteristics

An input source is characterized by:

- Size of calling population
- Behaviour of the arrivals
- Pattern of arrival of customers at the system.

An input source need not be homogeneous population but may consist of several sub-populations. For example, patients arriving at OPD of a hospital are usually of three categories: walk-in patients, patients with appointments and emergency patients. Each patient class places different demands on service facility, but the waiting expectations of each category differ significantly.

Sue of Calling Population : The size of calling population whether it is homogeneous or consist of several sub-populations is considered either to be finite or infinite. When the number of potential arrivals is dependent on the number of customers already in the system (those being serviced plus those in queue), the calling population is *limited or finite.* An example of a finite calling population is a factory having only four machines which often require repair/service and two of them (say) are in working condition. Then at any point in time, there are only two machines which could possibly require service.

Alternately, if a new customer's arrival is independent of the number of customers already in the system, the calling population is considered *unlimited or infinite.* Examples of infinite population include customers arriving at a bank or super market, students arriving to get admission at the university, cars arriving at a highway petrol pump, etc.

Behaviour of Arrivals : If a customer, on arriving at the service system stays in the system until served, no matter how much he has to wait for service is called a *patient customer.* Machines arrived at the maintenance shop in a plant are examples of patient customers. Whereas the customer, who waits for a certain time in the queue and leaves the service system without getting service due to certain reasons such as a long queue in front of him is called an *impatient customer.* For example, a customer who has just arrived at a grocery store and finds that the salesmen are busy in serving the customers already in the system, will either wait for service till his patience is exhausted or estimates that his waiting time may be excessive and so leaves immediately to seek service elsewhere.

(i) Customers even before joining the queue get discouraged by seeing the number of customers already in service system or estimating the excessive waiting time for desired service, so they balk.

(ii) Other customers after joining the queue, wait for sometime and leave the service system due to intolerable delay, so they renege.

(iii) Customers who move from one queue to another hoping to receive service more quickly are said to be jockeying.

Pattern of arrivals The arrival process (or pattern) of customers to the service system is classified into two categories: static and dynamic. These two are further classified based on the nature of arrival rate and the control which can be exercised on the arrival process.

In static arrival process, the control depends on the nature of arrival rate (random or constant): *random* arrivals are either at a constant rate or varying with time. Thus to analyse the queuing system, it is necessary to attempt to describe the probability distribution of arrivals. From such distributions we obtain average time between successive arrivals, also called *inter-arrival time* (time between two consecutive arrivals and the average arrival rate (*i.e.*, number of customers arriving per unit of time at the service system).

The dynamic arrival process is controlled by both service facility and customers. The service facility adjusts' its capacity to match changes in the demand intensity, by either varying the staffing levels at different timings of service, varying service charges (such as telephone call charges at different hours of the day or week) at different timings, or allowing entry with appointments.

The variation in the demand intensity also affects the customer's behaviour. They either balk or renege from the service system when confronted with a long or slow moving waiting line.

The arrival time distribution can be approximated by one of the following probability distributions:

(i) Poisson distribution

(ii) Exponential distribution

(iii) Erlang distribution.

The Poisson distribution is a discrete probability distribution of the number of customers (or events) arriving in some time interval. Let us consider a Poisson process involving the number of arrivals n over a time period t. If λ is the expected (or average) number of arrivals per unit time, then expected number of arrivals during a time interval t will be λt. Then Poisson probability mass function is

$$P(x = n \mid P_n = \lambda t) = \frac{(\lambda t)^n e^{-\lambda t}}{n!}, \quad n = 0, 1, 2...$$

Now, if we are interested in the inter-arrival time probability distribution, then in the time interval from 0 to t, the probability of no arrival is given by

$$P(x = 0 \mid P_n = \lambda t) \frac{(\lambda t)^0 e^{-\lambda t}}{0!} = e^{-\lambda t}.$$

Let us define the random variable T as the time between successive arrivals. Since a customer can arrive at any time, T must be a continuous random variable. The probability of no arrival in the time interval from 0 to t will be equal to the probability that T exceeds t, so we have

$$P(T > t) = P(x = 0 \mid P_n = \lambda t) = e^{-\lambda t}.$$

The probability that there is an arrival in the time interval from 0 to t is given by

$P(T \leq t) = 1 - P(T > t) = 1 - e^{-\lambda t}; t \geq 0.$

This probability is also called the cumulative probability distribution function, F(t) of T. Also the distribution of the random variable T is referred to as the *exponential distribution*. Its probability density function (pdf) is

$$f(t) = \begin{cases} \lambda e^{-\lambda t}; & t \geq 0 \\ 0 & ; t < 0 \end{cases}$$

In many practical situations, the inter-arrival time is approximated by an exponential distribution (sometimes called the negative exponential distribution).

The mean of exponential distribution is the expected or average time between arrivals E(T). Thus, with λ arrivals per unit time, E(T) = 1/λ.

Hence, from above discussion we may conclude that the Poisson distribution of arrivals with arrival rate A, is equal to the negative exponential distribution of inter-arrival times with mean inter-arrival time 1/λ.

The probability density function of exponential distribution can also be used to compute the probability that the next customer arrives within time, T of the previous arrival. This means that if a customer has already arrived at the service system, then probability of arriving for the next customer can be determined by using this equation. For example, if λ = 1.2 and t = 1.5, λt = 1.8, so

$$F(1.5) = P(T \leq 15) = 1 - (x = 0 \mid P_n = 1.8)$$
$$= 1 - 0.165 = 0.835.$$

Also, if customers arrive at a service system at an average rate of 24 customers per hour, and a customer

has already arrived, then probability of another customer arriving in the next 5 minutes (*i.e.*, t = 1/12 hr) is:

$$P(\text{inter-arrival time} \leq 1/12) = 1 - e^{-24(1/12)}$$
$$= 1 - e^{-2} = 1 - 0.1353 = 0.8647.$$

Further, if λ = 24 customers per hour, t = 1/12 hour and λt = 24 × (1/12) = 2, the probability of n = 2 customers arriving within the next 10 minutes is:

$$P(x = 2 \mid P_n = 2) = \frac{e^{-24(1/12)} (24/12)^2}{2!} = 0.27.$$

Queuing Process

The queuing process refers to the number of queues, and their respective lengths. The number of queues depend upon the layout of a service system. Thus there may be a *single queue* or *multiple queues*. Certain service systems adopt *'take-a-number'* policy to avoid a formal queue to form; the length (or size) of the queue depends upon the operational situation such as physical space, legal restrictions, and attitude of the customers.

In certain cases, a service system is unable to accommodate more than the required number of customers at a time. No further customers are allowed to enter until space becomes available to accommodate new customers. Such type of situations are referred to as *finite (or limited) source queue*. Examples of finite source queues are cinema halls, restaurants, etc. On the other hand, if a service system is able to accommodate any number of customers at a time, then it is referred to as *infinite (or unlimited) source queue*. For example, in a sales department where the customer orders are received, there is no restriction on the number of orders that can come in so that a queue of any size can form.

In many other situations, when arriving customers experience long queue(s) in front of a service facility, they often do not enter the service system even though additional waiting space is available. The queue length in such cases depends upon the *attitude of the customers*. For example, when a motorist finds that there are many vehicles waiting at the petrol station, in most of the cases he does not stop at this station and seeks service elsewhere.

In some finite source queue systems, the maximum permissible queue is of zero length, *i.e.*, no queue is allowed to form. For example, when a parking space (service facility) cannot accommodate additional incoming vehicles (customers), the motorists are diverted elsewhere.

A multiple queues system at a service system can also be finite or infinite. But this has certain advantages such as:

1. The service provided to the customers can be differentiated by adopting different service rules called *queue discipline*.
2. Division of manpower is possible.
3. Customer has the option of joining any queue and can also switch to the end of any other queue.
4. Balking behaviour of the customers can be controlled.

Queue Discipline

The queue discipline is the order or manner in which customers from the queue are selected for service. There are a number of ways in which customers in the queue are served. Some of these are:

(a) Static queue disciplines are based on the individual customer's status in the queue. Few of such disciplines are:

(i) If the customers are served in the order of their arrival, then this is known as the *first-come, first-served (FCFS)* service discipline. Prepaid taxi queue at airports where a taxi is engaged on a 'first-come, first-served' basis is an example of this discipline.

(ii) Other discipline which is also common in use is *last-come, first-served (LCFS)*. This discipline is practised in most cargo handling situations where the last item loaded is removed first. Another example is the production process where items arrive at a workplace and are stacked one on top of the other. Item on the top of the stack is taken first for processing which is the last one to have arrived for service.

(b) Dynamic queue disciplines are based on the individual customer attributes in the queue. Few of such disciplines are:

(i) *Service in Random Order (SIRO)* : Under this rule customers are selected for service at random irrespective of their arrivals in the service system.

(ii) *Priority Service* : Under this rule customers are grouped in priority classes on the basis of some attributes such as service lime or urgency, and FCFS rule is used within each class to provide service. The payment of telephone or electricity bills by cheque or cash are examples of this discipline.

(iii) *Pre-Emptive Priority* : Under this rule, the highest priority customer is allowed to enter into the service immediately after entering into the system even if a customer with lower priority is already in service. That is, lower priority customer's service is interrupted (pre-empted) to start service for a special customer. This interrupted service is resumed again after the highest priority customer is served or according to the pre-emptive rules.

(iv) *Non-Pre-Emptive Priority* : In this case highest priority customer goes ahead in the queue, but service is started immediately on completion of the current service.

APPLICATIONS OF QUEUING MODELS

Waiting line or queuing theory has been applied to a wide variety of business situations. All situations where customers are involved such as restaurants, cafeterias, departmental stores, cinema halls, banks, post-offices, petrol pumps, airline counters, patients in clinics, etc., are likely to have waiting lines. Generally, the customer expects a certain level of service, whereas the from providing service facility tries to keep the costs minimum while providing the required service.

Service Channels

The queuing system may have a single service channel. Arriving customers may form one line and gel serviced, as in a doctor's clinic. The system may have a number of service channels, which may be arranged in parallel or in series or a complex combination of both. In case of parallel channels, several customers may be serviced simultaneously, as in a barber shop. For series channels, a customer must pass successively through all the channels before service is completed, *e.g.*, a product undergoing different processes over different machines. A queuing model is called *one server model*, when the system has one server only and a *multi-server model* when the system has a number of parallel channels each with one server.

Sometimes, several service channels may feed into one subsequent service channel; for example, several ticket booths in a theatre may send all the ticket holders to a single ticket collector at the entrance of the theatre. On the other hand, sometimes, a single service channel may disperse customers among several channels that come after it; for example, an enquiry clerk in an office.

Service Discipline

Service discipline or order of service is the rule by which customers are selected from the queue for service. The most common discipline is 'first come, first served', according to which the customers are served in the order of their arrival, *e.g.*, cinema ticket windows, railway stations, tanks, etc. The other discipline is 'last come, first served', is in a big godown, where the items arriving last are taken out first. Still Other disciplines include 'random' and 'priority'.

'Priority' is said to occur when an arriving customer is chosen for service ahead of some Other customers already in the queue. A unit (customer) is said 10 have *'pre-emptive'* priority if it not merely goes to the head of the queue but displaces any unit already being served when it arrives. Provided that the order of service is not related to service time, it does not affect the queue length or average wailing time but it does affect the time an individual customer has to wail. The service discipline, therefore, affects the derivation of equations used for analysis. In this text only the most common service discipline 'first come, first served' will be assumed for further discussion.

Maximum Number of Customers Allowed in the System

Maximum number of customers in the system can be either finite or infinite. In some facilities, only a limited number of customers are allowed in the system and new arriving customers are not allowed to join the system unless the number becomes less than the limiting value.

MODELS FOR ARRIVAL AND SERVICE TIMES

Generally, arrivals do not occur at fixed regular intervals of times but tend to be clustered or scattered in some fashion. A *Poisson distribution* is a discrete probability distribution which predicts the number of arrivals in a given time. The Poisson distribution involves the probability of occurrence of an arrival. Poisson assumption is quite restrictive in some cases. It assumes that arrivals are random and independent of all other operating conditions. The mean arrival rate (*i.e.,* the number of arrivals per unit of time) A, is assumed to be constant even time and is independent of the number of units already serviced, queue length or any other random property of the queue.

Since the mean arrival rate is constant over time, it follows that the probability of an arrival between time t and t + dt is λ. dt.

Thus probability of an arrival in time dt = λ.dt. ...(1)

The following characteristics of Poisson distribution are written here without proof:

Probability of n arrivals in time $t = \frac{(\lambda t)^n . e^{-\lambda t*}}{n!}$. ...(2)

Probability density function of inier-arrival time (time interval between two consecutive arrivals)

$$= \lambda . e^{-\lambda t}. \qquad ...(3)$$

Finally, Poisson distribution assumes that the time period dt is very small so that $(dt)^2$, $(dt)^3$, etc. → 0 and can be ignored.

Service time is the time required for completion of a service *i.e.,* it is the time interval between beginning of a service and its completion. The mean service rate is the number of customers served per unit of time (assuming the service to be continuous through the entire time unit), while the average service time 1/μ is the time required to serve one customer. The most common type of distribution used for service times is exponential distribution. It involves the probability of completion of a service. It should be noted that Poisson distribution cannot be applied to servicing because of the possibility of the service facility remaining idle for some time. Poisson distribution assumes fixed time interval of continuous servicing, which can never be assured in all services.

Mean service rate μ is also assumed to be constant over time and independent of number of units already serviced, queue length or any other random property of the system. Thus probability that a service is completed between t and t + dt, provided that the service is continuous

$$= \mu dt. \qquad ...(4))$$

Under the condition of continuous service, the following characteristics of exponential distribution are written, without proof:

Probability of n complete services in time $t = \frac{(\mu t)^n . e^{-\mu t}}{n!}$.

Probability density function (p.d.f.) of inter-service time, *i.e.*, time between two consecutive services = $\mu . e^{-\mu t}$.

Kendall's Notation for Representing Queuing Models

D.G. Kendall (1953) and later A. Lee (1966) introduced useful notation for queuing models. The complete notation can he expressed as

SOLVED EXAMPLES

Example 1:

A bank has two tellers working on saving accounts. The first teller handles withdrawals only while the second teller handles deposits only. It has been found that the service time distribution for the deposits and withdrawals both is exponential with mean service time 3 minutes per customer. Depositors are found to arrive in a Poisson fashion throughout the day with mean arrival rate 15 per hour. Withdrawers also arrive in a Poisson fashion with mean arrival rate 14 per hour.

(a) *What would be the effect on the average waiting time for depositors and withdrawers if each teller could handle both withdrawals and deposits?*

(b) *What would be the effect it this could only be accomplished by increasing the service time to 3.5 minutes?*

Solution:

Here we have, $\mu = \frac{6.(10/3)^3}{3!(24-20)^2}$/minute = 20/hour,

λ_1 = 6/hour, λ_2 = 14/hour.

Average waiting time for depositors,

$$W_q = \frac{1}{\mu} \cdot \frac{\lambda_1}{\mu - \lambda_1} = \frac{1}{20} \times \frac{16}{20-16} = \frac{1}{5} \text{ hours} = 12 \text{ minutes.}$$

Average waiting time for withdrawers,

$$W_{q2} = \frac{1}{\mu} \cdot \frac{\lambda}{\mu - \lambda_2} = \frac{1}{20} \times \frac{14}{20-14} = \frac{7}{60} \text{ hours} = 7 \text{ minutes.}$$

(a) Here, $\lambda = \lambda_1 + \lambda_2$ = 16 + 14 = 30/hour,

$$\mu = 20/\text{hour},\ \lambda/\mu = 3/2,\ c = 2.$$

$$\therefore \quad p_0 = \frac{1}{\sum_{n=0}^{c-1}\frac{(\lambda/\mu)^n}{n!} + \frac{(\lambda/\mu)^c}{c!}\cdot\frac{c\mu}{c\mu-\lambda}}$$

$$= \frac{1}{\sum_{n=0}^{1}\frac{(3/2)^n}{n!} + \frac{(3/2)^2}{2!}\cdot\frac{40}{40-30}}$$

$$= \frac{1}{(1+3/2)+\frac{9}{4\times 2}\times\frac{40}{10}} = \frac{1}{7}$$

$\therefore$ Average waiting time,

$$W_q = \frac{\mu.(\lambda/\mu)^c}{(c-1)!(c\mu-\lambda)^2}\cdot p_0 = \frac{20.(3/2)^2}{1!(40-30)^2}\times\frac{1}{7}$$

$$= \frac{9}{140}\text{ hours} = \frac{27}{7}\text{ minutes} = 3.86\text{ minutes.}$$

(b) When the service time is 3.5 minutes,

$$\text{service rate } \mu = \frac{1}{3.5}/\text{minute} = \frac{60}{3.5} = \frac{120}{7}/\text{hour},$$

$$\lambda = 30/\text{hour},\ \frac{\lambda}{\mu} = \frac{7}{4},\ c = 2.$$

$$\therefore \quad p_0 = \frac{1}{\sum_{n=0}^{c-1}\frac{(\lambda/\mu)^n}{n!} + \frac{(\lambda/\mu)^c}{c!}\cdot\frac{c\mu}{c\mu-\lambda}}$$

$$= \frac{1}{\sum_{n=0}^{1}\frac{(7/4)^n}{n!} + \frac{(7/4)^2}{2!}\cdot\left(\frac{240/7}{240/7-30}\right)}$$

$$= \frac{1}{\left(1+\frac{7}{4}\right)+\frac{49}{2\times 16}\times 8} = \frac{1}{15}.$$

$\therefore$ Average waiting time,

$$W_q = \frac{\mu(\lambda/\mu)^c}{(c-1)!(c\mu-\lambda)^2}\cdot p_0$$

$$= \frac{120/7(7/4)^2}{1!(240/70-30)^2}\cdot\frac{1}{15}$$

$$= \frac{343}{1800} \text{ hours} = 11.43 \text{ minutes.}$$

Example 2:

At present a servicing department provides answers through one channel, which on average can deal with 24 enquiries/hour at a cost of f3 per enquiry. Increasingly the customers are complaining that they have to wait for a long time and the department is considering alternative arrangements. There is either a two-channel system costing f 100/hour and service rate of 15/hour in each, or a three channel system costing f125 per hour and service rate of 10/hour in each. Customers arrive at the rat of 20/hour. Average time a customer is in this system

$$= \frac{(\rho k)^k}{k!(1-\rho)^2).k\mu} p_0 + \frac{1}{\mu},$$

where $$p_0 = \frac{k!(1-\rho)}{(\rho k)^k + k!(1-\rho)\left\{\sum_{n=0}^{k-1} \frac{(\rho k)^n}{n!}\right\}}.$$

Your are required to calculate

(a) *The average time a customer is in the system under the present arrangement,*

(b) *The extracharges per enquiry that would need to be made to recover the extra cost of each of the two arrangements proposed,*

(c) *The implied value of customer's time per hour if they agree to pay the extra cos to f the two-channel system.*

Solution:

Here, $\lambda = 20$/hour, $\mu = 24$/hour.

(a) Average time a customer spends in the system,

$$W_s = \frac{1}{\mu - \lambda} = \frac{1}{24-20} = \frac{1}{4} \text{ hour} = 15 \text{ minutes.}$$

(b) Present cost = f 3 × 24 = f 72/hour.

(i) Cost of 2-service channel system = f 100/hour.

∴ Extra charges/enquiry = f (100 – 72)/20 = f 1.40.

(ii) Cost of 3-service channel system = f 125/hour.

∴ Extra charges/enquiry = f (125 – 72)/20 = f 2.65.

(c) Here $\mu = 5$/hour, k = 2.

$$\therefore \quad \rho = \frac{\lambda}{k\mu} = \frac{20}{2\times15} = \frac{2}{3}.$$

$$\therefore \quad p_0 = \frac{2!\left(1-\frac{2}{3}\right)}{\left(\frac{2}{3}\times2\right)^2 + 2!\left(1-\frac{2}{3}\right)\left\{\sum_{n=0}^{1}\frac{(2/3\times2)^n}{n!}\right\}}$$

$$= \frac{2\times1/3}{\frac{16}{9}+2\times\frac{1}{3}\cdot\left\{1+\frac{1}{1!}\left(\frac{2}{3}\times2\right)^1\right\}}$$

$$= \frac{2/3}{\frac{16}{9}+\frac{2}{3}\left\{1+\frac{4}{3}\right\}} = \frac{2}{3}\times\frac{9}{30} = \frac{1}{5} = 0.2$$

Average time spent in the system,

$$W_s = \frac{\left(\frac{2}{3}\times2\right)^2}{2!\left(1-\frac{2}{3}\right)^2(2\times15)}\times0.2+\frac{1}{15}$$

$$= \frac{16}{9}\times\frac{1}{2}\times\frac{9}{30}\times0.2+\frac{1}{15} = \frac{1.8}{15}$$

= 0.12 hours = 7.2 minutes.

∴ Customer's time saved = 15 – 7.2 = 7.8 minutes.

∴ The implied value of customer's per hour

$$= f\ 1.40 \times \frac{60}{7.8}$$

= f 10.77.

Example 3:

A branch of Punjab National Bank has only one typist. Since the typing work varies in length (number of pages to be typed), the typing rate is randomly distributed approximating a Poisson distribution with mean service rate of 8 letters/hour. The letters arrive at a rate of 5/hour during the entire8-hour work day. If the waiting time cost of the letters is Rs. 1.50 per hour, find the average system time and the total lost time cost.

There is a possibility of either installing an additional typewriter of the same type or replacing the present one by a better and faster typewriter. The data are given below.

Table 1

	Service rate	*Daily rental*
Present typewriter	*8*	*2.50*
Proposed typewriter	*12*	*4.50*

Suggest the better alternative.

Solution:

Here, λ = 5/hour, μ = 8/hour.

Average system time with present typewriter,

$$W_s = \frac{1}{\mu - \lambda} = \frac{1}{8-5} = \frac{1}{3} \text{ hour.}$$

$\therefore$ Lost time cost/day = 8 × 5 × 1/3 × Rs. 150 = Rs. 20

$\therefore$ Total cost per day of the present typewriter

= rental cost + lost time cost

= Rs. 2.50 + Rs. 20.00 = Rs. 22.50.

For proposed typewriter:

Average system time, $W_s = \frac{1}{\mu - \lambda} = \frac{1}{12-5} = \frac{1}{7}$ hour.

Lost time cost per day = $(8 \times 5) \times \frac{1}{7} \times 1.50$ = Rs. $\frac{60}{7}$ = Rs. 8.57.

Total cost per day = rental cost + lost time cost

= Rs. 4.50 + Rs. 8.57 = Rs. 13.07.

To calculate the waiting time in the system for two machines, first let us find p_0.

$$p_0 = \frac{1}{\sum_{n=0}^{n=c-1} \frac{1}{n!}\left(\frac{\lambda}{\mu}\right)^n + \frac{1}{c!}\left(\frac{\lambda}{\mu}\right)^c \frac{c\mu}{c\mu - \lambda}}$$

$$= \frac{1}{\sum_{n=0}^{n=1} \frac{1}{n!}\left(\frac{5}{8}\right)^n + \frac{1}{2!}\left(\frac{5}{8}\right)^2 \cdot \frac{2\times 8}{2\times 8 - 5}}$$

$$= \frac{1}{\frac{1}{0!}\left(\frac{5}{8}\right)^0 + \frac{1}{1!}\left(\frac{5}{8}\right)^1 + \frac{1}{2!}\left(\frac{5}{8}\right)^2 \cdot \frac{16}{11}}$$

$$= \frac{1}{1+\frac{5}{8}+\frac{25}{58}} = \frac{1}{\frac{88+55+25}{88}} = \frac{88}{168} = \frac{11}{21}.$$

Expected waiting time in the system,

$$W_s = \frac{\mu(\lambda/\mu)^c p_0}{(c-1)!(c\mu-\lambda)^2} + \frac{1}{\mu}$$

$$= \frac{8\times(5/8)^2\times 11/21}{(2-1)!(2\times 8-5)^2} + \frac{1}{8}$$

$$= \frac{25/8\times 11/21}{11\times 11} + \frac{1}{8}$$

$$= \frac{25}{88\times 21} + \frac{1}{8} = \frac{32}{231}.$$

∴ Total cost per day = rental cost of two typewriters + cost of lost time

$$= \text{Rs. } \left(2\times 2.50 + 8\times 5\times \frac{32}{231}\times 1.50\right)$$

$$= \text{Rs. } \left(5+\frac{640}{77}\right) = \text{Rs. } (5 + 8.31) = \text{Rs. } 13.31.$$

∴ It is slightly less expensive to use a single proposed better (faster) machines.

Example 4:

A mechanic repairs four machines. The mean time between service requirements is 5 hours for each machine and forms an exponential distribution. The mean time is one hour and also follows the same distribution pattern. Machine downtime costs Rs. 25 per hour the mechanic costs Rs. 55 per day. Determine the following:

(a) *Probability that the service facility will be idle,*

(b) *Probability of various number of machines (0 through 4) to be out of order and being repaired,*

(c) *Expected number of machines waiting to be repaired, and being repaired,*

(d) *Expected downtime cost per day.*

Would it be economical engage two mechanics, each repairing only two machines?

Solution:

It is given that

$$\lambda = 1/5 = 0.2 \text{ machine/hour} \; ; \; \mu = 1 \text{ machine/hour}$$

$$M = 4 \text{ machines} \; ; \; \rho = \lambda/\mu = 0.2$$

(a) The probability that the system shall be idle (or empty) is

$$P_0 = \left[\sum_{n=0}^{M} \frac{M!}{(M-n)!}\left(\frac{\lambda}{\mu}\right)^n\right]^{-1} = \left[\sum_{n=0}^{4} \frac{4!}{(4-n)!}(0.2)^n\right]^{-1}$$

$$= \left[1 + \frac{4!}{3!}(0.2) + \frac{4!}{2!}(0.2)^2 + \frac{4!}{1!}(0.2)^3 + \frac{4!}{0!}(0.2)^4\right]^{-1}$$

$$= [1 + 4\,(0.2) + 4 \times 3\,(0.04) + (4 \times 3 \times 2)\,(0.008) + (4 \times 3 \times 2 \times 1)\,(0.00016)]^{-1}$$

$$= [1 + 0.8 + 0.48 + 0.192 + 0.000384]^{-1}$$

$$= (2.481)^{-1} = 0.030.$$

(b) Probability that there shall be various number of machines (0 through 4) in the system.

$$P_n = \frac{M!}{(M-n)!}\left(\frac{\lambda}{\mu}\right)^n P_0 \; ; \; n \le M$$

(c) The expected number of machines to be out of order and being repaired

$$L_s = M - \frac{\mu}{\lambda}(1 - P_0) = 4 - \frac{1}{0.2}(1 - 0.403)$$

$$= 4 - 2.985 = 1.015 \text{ machines}$$

Table 1 Calculation of P_n

n (1)	$\frac{M!}{(M-n)!}\left(\frac{\lambda}{\mu}\right)^n$ (2)	Probability* (3) = (2) × P_0
0	1.00	0.4030
1	0.80	0.3224
2	0.48	0.1934
3	0.19	0.0765
4	0.00	0.0000

(d) Expected time a machine will in queue to be repaired

$$W_q = \frac{1}{\mu}\left[\frac{M}{1 - P_0} - \frac{\lambda + \mu}{\lambda}\right] = \left[\frac{4}{1 - 0.403} - \frac{0.2 + 1}{0.2}\right]$$

$$= \frac{4}{0.597} - 6 = 0.70 \text{ hours or 42 minutes}$$

(e) If there are two mechanics each serving two machines, then M = 2, and therefore,

$$P_0 = \left[\sum_{n=0}^{2} \frac{M!}{(M-n)!}\left(\frac{\lambda}{\mu}\right)^n\right]^{-1} = [1 + 2\,(0.2) + 2 \times 1\,(0.2)^2] = 0.68$$

Cost Analysis : It is assumed that each mechanic with his two machines constitutes a separate system with no interaction. Thus,

(a) Expected number of machines in the system will be

$$L_S = M - \frac{\mu}{\lambda}(1 - P_0)$$

$$= 2 - \frac{1}{0.2}(1 - 0.68) = 0.4 \text{ machine}$$

The expected downtime of machines per day is

= Expected number of machines in the system × 8-hour day × Number of mechanics

= 0.4 × 8 × 2 = 6.4 hours/day

Total cost for hiring two mechanics will be

Total cost = Mechanics's cost + Downtime cost

= 2 × 55 + 6.4 × 25 = Rs. 270 per day

But the total cost with one mechanic is Rs. (55 + 200) = Rs. 255/day. Hence, it is not economical to engage two mechanics.

Example 5:

There are 5 machines, each of which when running suffers breakdown at an average rate of 2 per hour. There are 2 servicemen and only one man can work on a machine at a time. If n machines are out of order when n < 2 then (n – 2) of them wait until a serviceman is free. Once a serviceman starts work on a machine the time to complete the repair has an exponential distribution with mean of 5 minutes. Find the distribution of the number of machines out of action at a given time. Find also the average time an out-of-action machine has to spend waiting for the repairs to start.

Solution:

From the data of the problem, we have

λ = 2 per hour, μ = 60/5 per hour

M = 5 machines, and s = 2 servicemen

Then $$P_0 = \left[\sum_{n=0}^{2-1}\frac{5!}{(5-n)!n}\left(\frac{2}{12}\right)^n + \sum_{n=0}^{5}\frac{5!}{(5-n)!2!2^{n-1}}\left(\frac{2}{12}\right)^n\right]^{-1}$$

$$= \frac{648}{1493}$$

and $$P_n = \frac{5!}{(5-n)!n!}\left(\frac{2}{12}\right)^n\left(\frac{648}{1493}\right)P_0 \ ; 0 \le n < 2$$

$$\frac{5!}{(5-n)!2!2^{n-2}}\left(\frac{2}{12}\right)^n\left(\frac{648}{1493}\right)P_0 \ ; 2 \le n \le 5$$

(i) Average number of machines out-of-action at a given time

$$L_q = \sum_{n=2+1}^{5}(n-2)P_n = P_3 + 2P_4 + 3P_5$$

$$= \frac{165}{1493} = 0.110 \text{ hour.}$$

(ii) Average waiting time an out-of-action machine in the workshop

$$W_q = \frac{L_q}{\lambda_{eff}}$$

where $$\lambda_{eff} = 1\sum_{n=0}^{5}(5-n)P_n = \lambda\ (5P_0 + 4P_1 + 3P_2 + 2P_3 + P_4)$$

$$= 2\ (5P_0 + 3.333P_0 + 0.833P_0 + 0.023P_0)$$

$$= (2)\ \frac{648}{1493}\ (9.189) = 8 \text{ machines (approx.)}$$

Therefore $$W_q = \frac{0.110}{7.976} = 0.013 \text{ hour}$$

Example 6:

Solve example 35 assuming two equal sized docks.

Solution:

Here we have, λ = 3/hour, μ = 4/hour, λ/μ = 3/4, c = 2.

1. The probability that a truck has to wait for service,

$$p\ (n \ge c) = \frac{\mu.(\lambda/\mu)^c}{(c-1)!(c\mu-\lambda)}\cdot p_0$$

where $$p_0 = \frac{1}{\sum_{n=0}^{c-1}\frac{(\lambda/\mu)^n}{n!} + \frac{1}{c!}\left(\frac{\lambda}{\mu}\right)^c\cdot\frac{c\mu}{c\mu-\lambda}}$$

$$= \frac{1}{\sum_{n=0}^{n=1} \frac{1}{n!}(3/4)^n + \frac{1}{2!}(3/4)^2 \cdot \frac{2\times4}{2\times4-3}}$$

$$= \frac{1}{\frac{1}{0!}(3/4)^0 + \frac{1}{1!}(3/4)^1 + \frac{1}{2}\times\frac{9}{16}\times\frac{8}{5}}$$

$$= \frac{1}{1+\frac{3}{4}+\frac{9}{20}} = \frac{1}{\frac{20+15+9}{20}} = \frac{20}{44} = \frac{5}{11}.$$

$$\therefore \quad p\ (n \geq c) = \frac{4.(3/4)^2}{1!(4\times2-3)} = \frac{4.(3/4)^2}{1!(4\times2-3)}\times\frac{5}{11} = \frac{4\times9}{16\times5}\times\frac{5}{11}$$

$$= \frac{9}{11} = 0.205.$$

2. The waiting time of a truck that waits is,

   Expected waiting time of a truck, W_s

$$W_n = \frac{\text{Expected waiting time of a truck, } W_n}{\text{probability that a truck actually has to wait, } p_c}$$

$$\therefore \quad W_n = \frac{\mu.(\lambda/\mu)^c.p_0}{(c-1)!(c\mu-\lambda)^2}\times\frac{1}{p_c}$$

$$= \frac{4(3/4)^2}{1!(2\times4-3)^2}\times\frac{5}{11}\times\frac{44}{9} = 0.2.$$

3. Total expected waiting time of company trucks per day

   = Trucks/day × % of company trucks × Expected waiting time/truck

   = (3 × 8) × (0.40) × W_n

   = 24 × 0.40 × (p_c × W_n)

   = 24 × 0.40 × 9/44 × 0.2 = 0.393.

Example 7:

The average rate of arrivals at a service store is 30 customers per hour. At present there is one cashier who on an average attends 45 customers per hour. The store owner estimates that each extra minute of system process time per customer means an additional cost of Rs 1.5. An assistant can be provided to the cashier and in that case the service unit can deal with 85 customers per hour. Determine the rate (customers per hour) with which customers should be attended to minimize customer waiting cost and to justify the employment of an assistant.

Solution:

Given that,

λ = 30 customers/hour,

λ = 45 customers/hour and cost per additional customer service,

C_1 = Rs. 1.5. The cost of waiting per hour is as follows:

Cost of waiting = Expected number of customers in the system × Hourly

waiting cost

$$\frac{\lambda}{\mu - \lambda} \times 1.5 \times 60$$

$$\frac{30}{45 - 30} \times 1.5 \times 60 = \text{Rs. } 180/\text{hour.}$$

Therefore, C_2 = Rs. 180/hour.

The new value of service rate is obtained as

$$\mu = \lambda + X\sqrt{\frac{\lambda C_2}{C_1}} = 30 + \sqrt{\frac{30 \times 180}{1.5}}$$

$$= 90 \text{ customers/hour.}$$

Example 8:

In a factory cafeteria the customers have to pass through three counters. The customers buy coupons at the first counter, select and collect the snacks at the second counter and collect tea at the third. The server at each counter takes on an average 1.5 minutes although the distribution of service time is approximately exponential. If the arrival of customers to the cafeteria is approximately Poisson at an average rate of 6 per hour, calculate

(a) *The average time a customer spends waiting in the factorial,*

(b) *The average time of getting the service,*

(c) *The most probable time in getting the service.*

Solution:

Here we have, phases k = 3,

service time per phase = 1.5 minutes.

$\therefore$ Service time per customer = 1.5 × 3 = 4.5 minutes.

$$\therefore \quad \mu = \frac{1}{4.5} \text{ customers/minute} = \frac{40}{3} \text{ customers/hour,}$$

λ = 6 customers/hour.

(a) Average waiting time,

$$W_q = \frac{k+1}{2k}\cdot\frac{\lambda}{\mu}\cdot\frac{1}{\mu-\lambda} = \frac{3+1}{2\times 3}\cdot\frac{6\times 3}{40}\cdot\frac{1}{\frac{40}{3}-6}$$

$$= \frac{9}{220} \text{ hours} = \frac{27}{11} = 2.45 \text{ minutes.}$$

(b) Average time of getting the service *i.e.*, in collecting coupons, snacks, etc., is the mean of t when it is following the 3rd member of Erlang family.

$\therefore$ Average time spent $= \frac{1}{\mu} = \frac{3}{40}$ hours = 4.5 minutes.

(c) The most probable time spent in getting the service is the modal value of t for the 3rd member of Erlang family.

$\therefore$ Most probable time spent $= \frac{k-1}{\mu k} = \frac{3-1}{\frac{40}{3}\times} = \frac{1}{20}$ hour = 3 minutes.

Example 9:

Repairing a certain type of machine which breaks down in a given factory consists of 5 basic steps that must be performed sequentially. The time taken to perform each of the 5 steps is found to have an exponential distribution with mean 5 minutes and is independent of other steps. If these machines break down in a Poisson fashion at an average rate of two per hour and if there is only one repairman, what is the average idle time for each machine that has broken down?

Solution:

Here, no of phases k = 5,

service time per phase = 5 minutes.

$\therefore$ Service time per unit = 5 × 5 = 25 minutes.

$\therefore$ $\mu = \frac{1}{25}$ units/minute $= \frac{12}{5}$ units/hour.

λ = 2 units, hour.

Average idle time of the machine = Average time spent by the machine in the system,

$$W_s = \frac{k+1}{2k}\cdot\frac{\lambda}{\mu}\cdot\frac{1}{\mu-\lambda}+\frac{1}{\mu}$$

$$= \frac{5+1}{2\times 5}\times\frac{2\times 5}{12}\times\frac{1}{\frac{12}{5}-2}+\frac{5}{12}$$

$$= \frac{1}{2} \times \frac{5}{2} + \frac{5}{12} = \frac{20}{12} = \frac{5}{3} \text{ hours} = 100 \text{ minutes.}$$

Example 10:

A colliery working one shift per day uses a large number of locomotives which break down at random intervals, on average one failing per 8-hour shift. The filter carries out a standard maintenance schedule on each faulty locomotive. Each of the 5 main parts of this schedule takes an average of 1/2 hour but the time varies widely. How much time will be fitter have for other tasks and what is the average time a locomotive is out of service?

Solution:

Here we have, $k = 5$, $\lambda = \frac{1}{8}$/hour,

service time per part $= \frac{1}{2}$ hour.

$\therefore$ Service time per locomotive $= \frac{5}{2}$ hours.

$\therefore$ $\mu = \frac{5}{2}$/hour.

Fraction of time the fitter will have for other tasks = Fraction of time for which the fitter is idle

$$= 1 - \frac{\lambda}{\mu} = 1 - \frac{1/8}{2/5} = 1 - \frac{5}{16} = \frac{11}{16}.$$

$\therefore$ Time the fitter will have for other task in a day

$$= 11/16 \times 8 = 5.5 \text{ hours.}$$

Average time a locomotive is out of service

= Average time spent by the locomotive in the system

$$= \frac{k+1}{2k} \cdot \frac{\lambda}{\mu} \cdot \frac{1}{\mu - \lambda} + \frac{1}{\mu}$$

$$= \frac{5+1}{2 \times 5} \times \frac{1/8}{2/5} \cdot \frac{1}{\frac{2}{5} - \frac{1}{8}} + \frac{5}{2}$$

$$= \frac{5+1}{2 \times 5} \times \frac{1/8}{2/5} \cdot \frac{1}{\frac{2}{5} - \frac{1}{8}} + \frac{5}{2} = 3.18 \text{ hours.}$$

Example 11:

Let there be an automobile inspection situation with three inspection stalls. Assume that cars wait in such a way that when wall becomes vacant, the car at the head of the line pulls up to it. The station can accommodate almost four cars waiting (seven in station) at one time. The arrival pattern is Poisson with a mean of one car every minute during the peak hours. The service time is exponential with mean of 6 minutes. Find the average number of customers in the system during peak hours, the average waiting time and the average number per hour that cannot enter the station because of full capacity.

Solution:

From the data of the problem, we have

λ = 1 car per minute; μ = 1/6 car per minute; s = 3, N = 7.

Then traffic intensity, $\rho = \lambda/\mu = 6$

Therefore, $$P_0 = \left[\sum_{n=0}^{3-1} \frac{1}{n!}(6)^n + \sum_{n=3}^{7} \frac{(6)^n}{3^{n-3}3!}\right]^{-1} = \frac{1}{1141}$$

(a) Expected number of customers in the queue

$$L_q = \frac{(3\times 6)^3.6}{3!(-5)^2} \cdot \left(\frac{1}{1141}\right)[1-(6)^5-(-5)(5)(6)^4] = 3.09 \text{ cars}$$

(b) Expected number of customers in the systems

$$L_s = 3.09 + 3 - P_0 \sum_{n=0}^{2} \frac{(3-n)}{n!}(6)^n = 6.06 \text{ cars}$$

(c) Expected waiting time in the system

$$W_s = \frac{6.06}{1(1-P_7)} = \frac{6.06}{1-\frac{(6)^7}{3!3^4}\times\left(\frac{1}{1141}\right)} = 12.3 \text{ minutes}$$

since $$P_n = \frac{1}{s!s^{n-s}}\left(\frac{\lambda}{\mu}\right)^n P_0 \text{ for } s < n \leq N$$

(d) Expected number of cars per hour that cannot enter the station

$$60\,\lambda\,P_N = 60.1.P_7 = 60.\,\frac{(6)^7}{3!3^4}\left(\frac{1}{1141}\right)$$

$$= 30.4 \text{ cars per hour.}$$

Example 12:

An airline maintenance base has facilities only for overhauling one airline engine at a time. Hence, to return the airplanes into use at the earliest, the policy is to stagger the overhauling of the 4 engines of each airplane. In other words, only one engine is overhauled each time an airplane comes into the shop. Under this policy, airplanes have arrivals according to a Poisson process at a mean rate of 1/day. The time required for an engine overhaul has an exponential distribution with a mean of 1/2 day.

A proposal has been made to change the policy so as to overhaul all four engines consecutively each time an aeroplane comes into the shop. It is pointed out that although this will quadruple the expected service time, each plane would need to come into the shop only one-fourth time as often. Compare the two alternatives on a meaningul basis.

Solution:

The two alternatives will be compared on the basis of the waiting time cost of the airplanes requiring overhauling.

First alternative: This is M/M/1 : FCFS/∞/∞ queuing system.

∴ Expected number of airplanes in the system,

$$W_s = \frac{\lambda}{\mu - \lambda} = \frac{1}{2-1} = 1.$$

Second alternative: This is M/E/1 : FCFS/∞/∞ quacking system.

Here, $l = \frac{1}{4}$ airplanes/day,

no, of phases, k = 4,

mean service time per airplane $= \frac{1}{2} \times 4 = 2$ days.

∴ Mean service rate, $\mu = \frac{1}{2}$ airplane per day.

∴ Expected number of airplanes in the system,

$$L_s = \frac{k+1}{2k}\frac{\lambda}{\mu}\cdot\frac{\lambda}{\mu-\lambda}+\frac{\lambda}{\mu}$$

$$= \frac{5}{8}\times\frac{1/4}{1/2}\times\frac{1/4}{1/2-1/4}+\frac{1/4}{1/2}$$

$$= \frac{5}{8}\times\frac{1}{2}\times 1+\frac{1}{2} = \frac{13}{16}.$$

Comparing the two, the second alternative results in smaller no, of planes requiring overhauling and hence lowers the waiting cos and should, therefore, be adopted.

Example 13:

In a car manufacturing plant, a loading crane takes exactly 10 minutes to load a car into a wagon and again come back to the position of loading another car. If the arrivals of cars is a Poisson stream at an average of one every 20 minutes, calculate the average waiting time of a car.

Solution:

Here, service rate is constant, being equal to 6/hour. Therefore k = ∞ and the average time of a car is given by

$$W_s = \frac{1}{2}\cdot\frac{\lambda}{\mu}\cdot\frac{1}{\mu-\lambda}$$

$$= \frac{1}{2}\cdot\frac{3}{6}\cdot\frac{1}{1-3} = \frac{1}{12} \text{ hour} = 5 \text{ minutes.}$$

Example 14:

The milk plant at a city distributes its products by trucks, loaded at the loading dock. It has it sown fleet of trucks of a private transport company. This transport company has complained that sometime trucks have to wait in line and thus the company loses money paid for a truck and driver that is only waiting. The company has asked the milk plant management either to go in for a second loading dock or discount prices equivalent to the waiting time. The following data are available.

Average arrival rate (all trucks) = 3 per hour.

average service rate = 4 per hour.

The transport company has provided 40% of the total number of trucks. Assuming that these rates are random according to Poisson distribution, determine

1. *The probability that a truck has to wait.*
2. *The waiting time of a truck that waits.*
3. *The expected waiting time of company trucks per day.*

Solution:

1. The probability that a truck has to wait for service = utilization factor

$$= \rho = \frac{\lambda}{\mu} = 3/4 = 0.75.$$

2. The waiting time of a tank that waits

$$= W_s = \frac{1}{\mu-\lambda} = \frac{1}{4-3} = \frac{1}{1} = \text{hour.}$$

Total expected waiting time of company trucks per day = Trucks day × % of company trucks × expected waiting time per truck

$$= (3 \times 8) \times (0.40) \times \frac{\lambda}{\mu(\mu-\lambda)}$$

$$= 24 \times 0.4 \times \frac{3}{4(4-3)} = 7.2 \text{ hours/day.}$$

Example 15:

A self-service store exploys one cashier at its counter. Nine customers arrive on an average every 5 minutes while the coheir can serve 10 customers in 5 minutes. Assuming Poisson distribution for arrival rate and exponential distribution for service rate, find

1. *Average number of customers in the system.*
2. *Average number of customers in queue or average queue length.*
3. *Average time a customer spends in the system.*
4. *Average time a customer waits before being served.*

Solution:

Arrival rate $\lambda = 9/5 = 1.8$ customers/minute,

service rate $\mu = 10/5 = 2$ customers/minute.

1. Average number of customers in the system

$$L_s = \frac{\lambda}{\mu - \lambda} = \frac{1.8}{2-1.8} = 9.$$

2. Average number of customers in the queue

$$\text{(Equation 10.15) } L_s = \frac{\lambda^2}{\mu(\mu-\lambda)} = \frac{\lambda}{\mu} \cdot \frac{\lambda}{(\mu-\lambda)}$$

$$= \frac{1.8}{2} \times \frac{1.8}{2-1.8} = 8.1$$

3. Average time a customer spends in the system,

$$W_s = \frac{1}{\mu-\lambda} = \frac{1}{2-1.8} = 5 \text{ minutes.}$$

4. Average time a customer spends in the queue,

$$W_s = \frac{\lambda}{\mu}\left(\frac{1}{\mu-\lambda}\right) = \frac{1.8}{2}\left(\frac{1}{2-1.8}\right) = 4.5 \text{ minutes.}$$

Example 16:

A person repairing radios finds that the time spent on the radio sets has been exponential distribution with mean 20 minutes. If the radios are repaired in the order in which they come in and their arrival is approximately Poisson with an average rate of 15 for 8-hour day, what is the repairman's expected idle time each day? How many jobs are ahead of the average set just brought in?

Solution:

Arrival rate $\lambda = \frac{15}{8 \times 60} = \frac{1}{32}$ units/minute,

service rate $\mu = \frac{1}{20}$ units/minute.

Number of jobs ahead to the set brought in = Average number of jobs in the system,

$$L_s = \frac{\lambda}{\mu - \lambda} = \frac{1/32}{1/20 - 1/32} = \frac{5}{3}.$$

Number of hours for which the repairman remains busy in an 8 hour day

$$= 8 \times \frac{\lambda}{\mu} = 8 \times \frac{1/32}{1/20} = 8 \times \frac{20}{32} = 5 \text{ hours.}$$

$\therefore$ Time for which repairman remains idle in an 8-hour day

$$= 8 - 5 = 3 \text{ hours.}$$

Example 17:

Consider a situation in which the niche arrival rate is one customer every 4 minutes and the mean service time is 2½ minutes. If the wailing cost is Rs. 5 per unit per minute and the cost of servicing one unit is Rs. 4, find the minimum cost service rate.

Solution:

Here we have

$$\lambda = \frac{1}{4} = 0.25$$

$$\mu = \lambda \pm \sqrt{\frac{C_w}{C_f} . \lambda}$$

$$= 0.25 \pm \sqrt{\frac{5 \times 0.25}{4}} = 0.25 \pm \sqrt{0.3125} = 0.25 \pm 0.56$$

$\therefore \mu = 0.81$ units/minute. ($\because \mu = -0.31$ is not a feasible solution.)

Customer's Behaviour : The customer's behaviour is also very important in the study of queues. If a customer decides not to enter the queue since it is too long, he is said 10 have *balked.*

If a customer enters the queue, but after sometime loses patience and leaves it, he is said to have renewed. When there are two or more parallel queues and the customers move from one queue to the other, they are said to be jockeying.

Example 18:

Truck drivers who arrive to unload plastic materials for recycling currently wait an average of 15 minutes before unloading. The cost of driver and truck time wasted while in queue is valued Rs. 100 per hour. A new device is installed to process truck loads at a constant rate of 10 trucks per hour at a cost of Rs. 3 per truck unloaded. Trucks arrive according to a Poisson distribution at an average rate of 8 per hour. Suggest whether device should be put to use or not.

Solution:

Waiting cost before the use of new device

Waiting cost = Waiting time × Cost of waiting

= (15/60) × 100 = Rs. 25 per trip

Under new system λ = 8 per hour, and μ = 10 per hour

$$W_q = \frac{\lambda^2}{2\mu(\mu - \lambda)} = \frac{8}{2(10)(10-8)} = 1/5 \text{ hour} = 12 \text{ minutes}$$

Waiting cost per trip with new device = (1/5) × 100 = Rs. 20 per trip

Saving with new device = Current system – New system

= 25 – 20 = Rs. 5 per trip

New savings = (Savings – Cost) of installing new device

= (5 – 3) = Rs. 2 per trip.

Example 19:

A branch of Punjab National Bank has only one typist. Since the typing work varies in length (number of pages to be typed), the typing rate is randomly distributed approximating a Poisson distribution with mean service rate of 8 letters per hour. The letters arrive at a rate of 5 per hour during the entire 8-hour work day. If the typewriter is valued at Rs. 1.50 per hour, determine

1. *Equipment utilization.*
2. *The per cent time that an arriving letter has to wait.*
3. *Average system time.*
4. *Average cost due to waiting on the part of typewriter.*

Solution:

Arrival rate, λ = 5 per hour,

service rate, μ = 8 per hour,

1. Equipment utilization, $\rho = \frac{\lambda}{\mu} = \frac{5}{8} = 0.625$.
2. The per cent time an arriving letter has to wait
   = per cent time the typewriter remains busy
   = 62.5%
3. Average system time, $W_s = \frac{1}{\mu - \lambda} = \frac{1}{8-5} = \frac{1}{3}$ hr. = 20 minutes.
4. Average cost due to waiting on the part of the typewriter per day
   = 8 × (1 – 5/8) × Rs. 1.50 = Rs. 4.50.

Example 20:

Consider a single server queuing system with Poisson input, exponential service times. Suppose the mean arrival rate is 3 calling units per hour, the expected service time is 0.25 hour and the maximum permissible calling units in the system is two. Derive the steady-state probability distribution of the number of calling units in the system, and then calculate the expected number in the system.

Solution:

From the data of the problem, we have

$\lambda = 3$ units per hour; $\mu = 4$ units per hour, and $N = 2$

Then traffic intensity, $\rho = \lambda/\mu = 3/4 = 0.75$

The *steady-state probability distribution* of the number of n customers (calling units) in the system is

$$P_n = \frac{(1-\rho)\rho^n}{1-\rho^{N+1}};\ r \neq 1$$

$$= \frac{(1-0.75)(0.75)^n}{1-(0.75)^{2+1}} = (0.43)\ (0.75)^n$$

and $P_0 = \frac{(1-\rho)}{1-\rho^{N+1}} = \frac{1-0.75}{1-(0.75)^{2+1}} = \frac{0.25}{1-(0.75)^3} = 0.431$

The expected number of calling units in the system is given by

$$L_s = \sum_{n=1}^{N} nP_n = \sum_{n=1}^{N} n(0.43)(0.75)^n$$

$$= 0.43 \sum_{n=1}^{N} n(0.75)^n = 0.43\ \{(0.75) + 2\ (0.75)^2\} = 0.81.$$

Example 21:

In a large maintenance department, fitters draw parts from the parts-stores which is at present staffed by one store man. The maintenance foreman is concerned about the time spent by fitters getting parts and wants to know if the employment of a stores labourer would be worth while. On investigation it is found that

(a) a simple queue situation exists

(b) fitters cost Rs. 2.50 per hour,

(c) the storeman cost Rs. 2 per hour and can deal, on average, with 10 fitters per hour.

(d) a labourer could be employed at Rs. 1.75 per hour and would increase the service capacity of the stores to 12 per hour.

(e) on average 8 fitters visit the stores such hour.

Solution:

The problem can be solved by two methods:

Method I: Here, we calculate the average number of customers in the system before and after the labourer is employed and compare the reduction in the resulting quaking cost with the increase in service costs.

Without labourer:

λ = 8/hr, μ = 10/hr.

$\therefore$ Number of customers in the system, $L_s = \dfrac{\lambda}{\mu - \lambda} = \dfrac{8}{10-8} = 4.$

$\therefore$ Cost/hr = 4 × Rs. 2.50 = Rs. 10.

With labourer

λ = hr, μ = 12/hr.

$\therefore$ Number of customers in the system, $L_s = \dfrac{\lambda}{\mu - \lambda} = \dfrac{8}{12-8} = 2.$

$\therefore$ Cost/hr = cost of fitters per hour + cost of labourer per hour

= 2 × Rs. 2.50 + Rs. 1.75 = Rs. 6.75.

Since there is net saving of Rs. 3.25, it is recommended to employ the labourer.

Method II: Here, we calculate the average time spent by the customers (fitters) in the system before and after the employment of the labourer, and again compare the reduction in the resulting queuing cost with the increase in the service cost.

Without labourer

$\lambda = 8/\text{hr}, \mu = 10/\text{hr}.$

$\therefore$ Average time spent by a fitter in the system,

$$W_s = \frac{1}{\mu - \lambda} = \frac{1}{10-8} = 0.5 \text{ hr.}$$

$\therefore$ Cost/hr = 8 × 0.5 × Rs. 2.50 = Rs. 10.

With labourer

$\lambda = 8/\text{hr}, \mu = 12/\text{hr}.$

$\therefore$ Average time spent by a fitter in the system,

$$W_s = \frac{1}{\mu - \lambda} = \frac{1}{12-8} = \frac{1}{12-8}.$$

$\therefore$ Cost/hr = 8 × 1/4 × Rs. 2.50 + Rs. 1.75 = Rs. 6.75.

$\therefore$ It is recommended to employ the labourer.

Example 22(a):

On average 96 patients per 24 hour day require the service of an emergency clinic. Also on average, a patient requires 10 minutes of active attention. Assume that the facility can handle only one emergency at a time. Suppose that it costs the clinic Rs. 100 per patient treated to obtain an average servicing time of 10 minutes. And that each minute of decrease in this average time would cost Rs. 10 per patient treated. How much would have to be budgeted by the clinic to decrease the average size of the queue from $1^{1/3}$ patients to 1/2 patient?

Solution:

Here we have, $\lambda = \dfrac{96}{24} = 4$ patients/hr,

$$\mu = \frac{1}{10} \times 60 = 6 \text{ patients/hr.}$$

Average expected number of patients in the queue

$$= \frac{\lambda}{\mu} \cdot \frac{\lambda}{\mu - \lambda} = \frac{4}{6} \cdot \frac{4}{6-4}$$

or $$L_s = \frac{4}{6} \times \frac{4}{2} = \frac{4}{3} = 1\frac{1}{3}.$$

This number is to be reduced from $1\frac{1}{3}$ to $\frac{1}{2}(L'_s)$

Now $$L'_s = \frac{\lambda}{\mu'} \cdot \frac{\lambda}{\mu' - \lambda}.$$

$$\therefore \qquad \frac{1}{2} = \frac{4}{\mu'} \cdot \frac{4}{\mu'-4}$$

or $\qquad \mu'^2 - 4\mu' - 32 = 0$ or $(\mu' - 8)(\mu' + 4) = 0$

or $\qquad \mu' = 8$ patients/hr. ($\mu' = -4$ is illogical and hence neglected).

Average time required by each patient = 1/8hr = 15/2 minutes.

Decrease in the time required to attend a patient $= 10 - \frac{15}{2} = \frac{5}{2}$ minutes.

Therefore, the budget required for each patient = Rs. (100 + 5/2 × 10)

= Rs. 125.

Thus to decrease the size of the queue, the budget per patient should be increased from Rs. 100 to Rs. 125.

Example 22(b):

Workers come to tool store room to receive special tools (required by them) for accomplishing a particular project assigned to them. The average time between two arrivals is 60 seconds and the arrivals are assumed to be in Poisson distribution. The average service time (or the tool room attendant) is 40 seconds. Determine:

(a) average queue length,

(b) average length of non-empty queues,

(c) average number of workers in system including the worker being attended,

(d) mean waiting time of an arrival,

(e) average waiting time of an arrival (worker) who waits, and

(f) the type of policy to be established. In other words, determine whether to go in for an additional tool store room attendant which will minimize the combined cost of attendants' idle time and the cost of workers' waiting time. Assume the charges of a skilled worker Rs. 4 per hour and that of store room attendant Rs. 0.75 per hour.

Solution:

Here we have, λ = 1/60 per second = 1 per minute,

μ = 1/40 per second = 1.5 per minute.

(a) Average queue length,

$$L_s = \frac{\lambda}{\mu} \cdot \frac{\lambda}{\mu - \lambda} = \frac{1}{1.5} \cdot \frac{1}{1.5 - 1} = \frac{1}{0.75} = \frac{4}{3} \text{ workers.}$$

(b) Average length of non-empty queue,

$$L_s = \frac{\mu}{\mu-\lambda} = \frac{1.5}{1.5-1} = 3 \text{ workers.}$$

(c) Average number of workers in the system,

$$L_s = \frac{\lambda}{\mu-\lambda} = \frac{1}{1.5-1} = 2 \text{ workers.}$$

(d) Mean waiting time of an arrival,

$$W_q = \frac{1}{\mu}\cdot\frac{\lambda}{\mu-\lambda} = \frac{1}{1.5}\times\frac{1}{1.5-1} = \frac{4}{3} \text{ minutes.}$$

(e) Average waiting time of an arrival who waits,

$$W_n = \frac{1}{\mu-\lambda} = \frac{1}{1.5-1} = 2 \text{ minutes.}$$

(f) Probability that the tool room attendant remains idle

$$p_0 = 1 - \frac{1}{1.5-1}$$

$$= 1 - \frac{1}{1.5} = \frac{1}{3}.$$

∴ Idle time cost of one attenent = 1/3 × 8 × Rs. 0.75 = Rs. 2/day.

Waiting time cost of workers

$$= W_q \times \text{no. of workers arriving/day} \times \text{cost of worker}$$

$$= \left(\frac{4}{3}\times\frac{1}{60}\right) \times (8 \times 60) \times \text{Rs. } 4$$

$$= \text{Rs. } \frac{128}{3} = \text{Rs. } 42.67\text{/day.}$$

As the waiting time cost is much higher than the idle time cost, it is justified to employ an additional tool room attendant.

Example 22(c):

Customers arrive at a one window drive-in bank according to a Poisson distribution with mean 10 per hour. Service time per customer is exponential with mean 5 minutes. The space in front of the window, including that for the serviced car can accommodate a maximum of three cars. Other cars can wait outside this space.

(a) *What is the probability that an arriving customer can drive directly to the space in front of the window?*

(b) *What is the probability that an arriving customer will have to wait outside the indicated space?*

(c) *How long is an arriving customers expected to wait before starting service?*

Solution:

Here, $\lambda = 10/\text{hour}$, $\mu = 60/5 = 12/\text{hour}$.

(a) The probability that an arriving customer can drive directly to the space in front of the window

$$= p_0 + p_1 + p_2$$

$$= \left(1-\frac{\lambda}{\mu}\right)\frac{\lambda}{\mu}\left(1-\frac{\lambda}{\mu}\right)+\left(\frac{\lambda}{\mu}\right)^2\left(1-\frac{\lambda}{\mu}\right)$$

$$= \left(1-\frac{\lambda}{\mu}\right)\left[+\frac{\lambda}{\mu}+\frac{\lambda^2}{\mu^2}\right]$$

$$= \left(1-\frac{10}{12}\right)\left[1+\frac{10}{12}+\frac{100}{144}\right] = 0.42.$$

(b) The probability that an arriving customer has to wait outside the indicated space $= 1 - 0.42 = 0.58$.

(c) Average waiting time of a customer in the queue

$$= \frac{1}{\mu}\cdot\frac{\lambda}{\mu-\lambda} = \frac{10}{12(12-10)} = 0.417 \text{ hours.}$$

Example 23:

A super market has two sales girls at the sales counters. If the service time for each customer is exponential with a mean of 4 minutes, and if people arrive in a Poisson fashion at the rate of 10 an hour, then calculate the

(a) probability of having to wait for service?

(b) expected percentage of idle time for each sales girl?

(c) if a customer has to wait, what is the expected length of his waiting time?

Solution:

From the data of the problem, we have

$\lambda = 1/6$ per minute; $\mu = 1/4$ per minute,

$s = 2$ and $\rho = \lambda/s\mu = 1/3$

Therefore, $$P_0 = \left[\sum_{n=0}^{2-1}\frac{1}{n!}(4/6)^n + \frac{1}{2!}(4/6)^2\frac{2.(1/4)}{\{2.(1/4)-(1/6)\}}\right]^{-1}$$

$$= \left(1+\frac{2}{3}+\frac{1}{3}\right)^{-1} = \frac{1}{2}$$

and $P_1 = (\lambda/\mu)\ P_0 = (4/6)\ (1/2) = (1/3)$

(a) Probability of having to wait for service

$$P_w\ (n \geq 2) = \frac{1}{2!}\left(\frac{4}{6}\right)^2 \cdot \frac{2(1/4)}{2(1/4)-(1/6)}\left(\frac{1}{2}\right) = \frac{1}{6}$$

(b) The fraction of time the servers are busy, $\rho = \lambda/s\mu = 1/3$. Therefore, the expected idle time for each sales girl is $(1 - 1/3) = 2/3 = 6/$ 7%.

(c) Expected waiting time for a customer in the system

$$W_s = W_q + \frac{1}{\mu} = \frac{L_q}{\lambda} + \frac{1}{\mu}$$

$$= \frac{1}{(s-1)!}\left(\frac{\lambda}{\mu}\right)^s \cdot \frac{\mu}{(s\mu-\lambda)^2} . P_0 + \frac{1}{\mu}$$

$$= \left(\frac{4}{6}\right)^2 \frac{1/4}{[(1/2)-(1/6)]^2} \times \frac{1}{2} + 4 = 4.5 \text{ minutes.}$$

Example 24:

A tax consulting firm has 4 service counter in its office to receive who have problems and complaints about their income, wealth and sales taxes. Arrivals average 80 persons in an 8-hour service day. Each tax adviser spends an irregular amount of time servicing the arrivals which have been found to have an exponential distribution. The average service time is 20 minutes. Calculate the average number of customers in the system, average number of customers waiting to be serviced, average time a customer spends in the system, and average waiting time for a customer. Calculate how many hours each week does a tax adviser spend performing his job. What is the probability that a customer has two wait before he gets service? What is the expected number of idle tax advisers at any specified time?

Solution:

Give that, $\lambda = 10$/hour; $\mu = 3$/hour, $s = 4$; and $\rho = \lambda/s\mu = 5/6$

(a) Probability of no customer in the system.

$$P_0 = \left[\sum_{n=0}^{s-1}\frac{1}{n!}\left(\frac{\lambda}{\mu}\right)^n + \frac{1}{s!}\left(\frac{\lambda}{\mu}\right)^s\left(\frac{s\mu}{s\mu-\lambda}\right)\right]^{-1}$$

$$= \left[\sum_{n=0}^{3} \frac{1}{n!}\left(\frac{10}{3}\right)^n + \frac{1}{4!}\left(\frac{10}{3}\right)^4 \frac{12}{12-10}\right]^{-1}$$

$$= \left[1 + \frac{10}{3} + \frac{1}{2}\left(\frac{10}{3}\right)^2 + \frac{1}{6}\left(\frac{10}{3}\right)^3 + \frac{1}{24}\left(\frac{10}{3}\right)^4 .6\right]^{-1} = 0.021$$

(b) Average number of customers in the system

$$L_s = L_q = \frac{\lambda}{\mu} = \left[\frac{1}{(s-1)!}\left(\frac{\lambda}{\mu}\right)^s \frac{\lambda\mu}{(s\mu-\lambda)^2}\right] P_0 + \frac{\lambda}{\mu}$$

$$= \left[\frac{1}{3!}\left(\frac{10}{3}\right)^4 \frac{30}{(12-10)^2}\right] \times 0.021 + \frac{10}{3} = 6.57$$

(c) Average number of customers waiting in the queue length),

$$L_q = L_s - \frac{\lambda}{\mu} = 6.57 - (10/3) = 3.24 \text{ customers}$$

(d) Average time a customer spends in the system

$$W_s = W_q + \frac{1}{\mu} = \frac{L_q}{\lambda} + \frac{1}{\mu}$$

$$= \frac{3.24}{10} + \frac{1}{3} = 0.657 \text{ hour or } 39.42 \text{ minutes}$$

(e) Average time a customer waits for service in the queue

$$W_q = \frac{L_q}{\lambda} = \frac{3.24}{10} = 0.324 \text{ hour or } 19.44 \text{ minutes}$$

(f) Time spent by a tax counsellor, *i.e.*, utilization factor,

$$\rho = \frac{\lambda}{s\mu} = \frac{5}{6} = 0.833 \text{ hour or } 50 \text{ minutes}$$

The expected time spent in servicing customers during an 8-hour day is 8 × 0.833 = 6.66 hours. Thus, on an average, a tax advisor is busy for 6.66 × (40/8) = 33.30 hours based on a 40 hours week.

(g) Probability that a customer has to wait

$$P_w (n \geq s) = \frac{1}{s!}\left(\frac{\lambda}{\mu}\right)^s \frac{s\mu}{s\mu - \lambda} . P_0$$

$$= \frac{1}{4!}\left(\frac{10}{3}\right)^4 \cdot \frac{4 \times 3}{4 \times 3 - 10} (0.021) = 0.622$$

(h) The expected number of idle advisers at any specified time can be obtained by adding the probability of 3 idle, 2 idle and 1 idle advisers. That is

Expected number of idle advisers = $4P_0 + 3P_1 + 2P_2 + P_3$

$= 4\ (0.021) + 3\ (0.070) + 2\ (0.118) + 0.131 = 0.661.$

This means, less than one (0.661) adviser is idle on an average at any instance of time.

Example 25:

A repairman is to be fired to repair machines which break down at an average rate of 3 per hour. The breakdowns follow Poisson distribution. Non-productive time of machine is considered to cost Rs. 16 per hour. Two repairmen have been interviewed: one is slow but cheap, while the other is fast but expensive. The slow repairman charges Rs. 8 per hour and the he services broken down machines at the rate 4 per hour. The fast repairman demands Rs. 10 per hour and he services at an average rate of 6 per hour. Which repairman should be hired? Assume an 8-hour working day.

Solution:

Here we have, λ = 3/hour,

idle time (down time) cost of the machine = Rs. 16/hour.

Slow repairman

μ = 4/hour.

Average number of units in the system $= \dfrac{\lambda}{\mu - \lambda} = \dfrac{3}{4-3} = 3.$

Machine hours lost in 8-hour day = 3 × 8 = 24 hours.

Total cost/day = cost idle machines + repairman charges

= Rs. (16 × 24 + 8 × 8)

= Rs. (384 + 64) = Rs. 448.

Fast repairman μ = 6/hour.

Average number of units in the system $= \dfrac{\lambda}{\mu - \lambda} = \dfrac{3}{6-3} = 1.$

Machine hours lost in 8-hour day = 1 × 8 = 8 hours.

Total cost per day = cost of idle machines + repairman charges

= Rs. (16 × 8 + 10 × 8)

= Rs. (128 + 80) = Rs. 208.

Example 26:

A shipping company has a single unloading dock with ships arriving in a Poisson fashion at an average rate of 3/day. The unloading time distribution

for a ship with n unloading crews is found to be exponential with average unloading time 1/2n days. The company has a large labour supply without regular working hours, and to avoid long waiting times, the company has a policy of using as many unloading crews as there are ships waiting in line or being unloaded. Find

(a) *The average number of unloading crews working at any time, and*

(b) *The probability that more than 4 crews will be needed.*

Solution:

Here we have, $\lambda = 3$ ships/day,

$$\mu = n\mu,$$

and mean service rate with one unloading crew, $\mu = 2$ ships/day.

(a) Expected (average) number of unloading crews

$$= L_s = \sum_{n=0}^{\infty} np_n = \sum_{n=0}^{\infty} n \cdot \sum \frac{\rho^n}{n!} e^{-\rho}$$

$$= e^{-\rho} \sum_{n=0}^{\infty} \frac{\rho.\rho^{n-1}}{(n-1)!} = \rho.e^{-\rho} \sum_{n=1}^{\infty} \frac{\rho^{n-1}}{(n-1)!}$$

$$= \rho.e^{-\rho}.\left(1+\rho+\frac{\rho^2}{2!}+\frac{\rho^3}{3!}+\ldots\right)$$

$$= \rho e^{-\rho}(e^{\rho}) = r = \frac{\lambda}{\mu} = \frac{3}{2} = 1.5$$

(b) Probability that more than 4 crews will be needed

= probability that there are at least 5 ships in the system

$$= \sum_{n=5}^{\infty} p_n = \sum_{n=0}^{\infty} p_n - \sum_{n=0}^{4} p_n$$

$$= 1 - (p_0 + p_1 + p_2 + p_3 + p_4)$$

$$= 1 - e^{-\rho}\left[1+\rho+\frac{\rho^2}{2!}+\frac{\rho^3}{3!}+\frac{\rho^4}{4!}\right]$$

$$= 1 - e^{-1.5}\left[1+1.5+\frac{1.5^2}{2!}+\frac{1.5^3}{3!}+\frac{1.5^4}{4!}\right] = 0.019.$$

Example 27:

A bank has two tellers working on savings accounts. The first teller handles withdrawals only. The second teller handles deposits only. It has been found that the service time distribution for the deposits and withdrawals both are exponential with mean service time 3 minutes per customer. Depositors are found to arrive in a Poisson fashion throughout the day with a mean arrival rate of 16 per hour. Withdraws also arrive in a Poisson fashion with a mean arrival rate of 14 per hour. What would be the effect on the average waiting time for depositors and withdrawers if each teller could handle both withdrawals and deposits. What would be the effect is this could only be accomplished by increasing the service time to 3.5 minutes?

Solution:

Initially there are two independent queuing systems: Withdrawers and Depositors, where arrivals follow Poisson distribution and service time follows exponential distribution.

For withdrawers: Given that, $\lambda = 14$/hour and $\mu = 3$/minute or 20/minutes

Average waiting time in queue,

$$W_q = \frac{\lambda}{\mu(\mu-\lambda)} = \frac{14}{20(20-14)} = \frac{7}{60} \text{ hour or 7 minutes}$$

For depositors: Give that, $\lambda = 16$/hour, and $\mu = 3$/minute or 20 hour

Average waiting time in queue,

$$W_q = \frac{\lambda}{\mu(\mu-\lambda)} = \frac{16}{20(20-16)} = \frac{1}{5} \text{ hour or 12 minutes}$$

Combined Case: In this case there will be a common queue with two servers (tellers). Thus, we have

$\lambda = 14 + 16 = 30$/hour; $\mu = 20$/hour; $s = 2$; $\rho = \lambda/s\mu = 3/4$.

$$\text{Now, } P_0 = \left[\sum_{n=0}^{s-1}\frac{1}{n!}\left(\frac{\lambda}{\mu}\right)^n + \frac{1}{s!}\left(\frac{\lambda}{\mu}\right)^s\left(\frac{s\mu}{s\mu-\lambda}\right)\right]^{-1}$$

$$= \left[\sum_{n=0}^{1}\frac{1}{n!}\left(\frac{3}{2}\right)^n + \frac{1}{2!}\left(\frac{3}{2}\right)^s\left(\frac{40}{40-30}\right)\right]^{-1}$$

$$= \left[1+\frac{3}{2}+\frac{1}{2}\left(\frac{9}{4}\right)\cdot 4\right]^{-1} = \frac{1}{7}$$

Average waiting time of arrivals in the queue

$$W_q = \frac{L_q}{\lambda} = \left[\frac{1}{(s-1)!}\left(\frac{\lambda}{\mu}\right)^s \frac{\mu}{(s\mu-\lambda)^2}\right]P_0$$

$$= \left(\frac{3}{2}\right)^2 \frac{20}{(40-30)^2}\times\frac{1}{7} = \frac{9}{140} \text{ hour or 3.86 minutes.}$$

Combined waiting time with increased service time, when λ = 30/hour, μ = 60/3.5 or 120/7 per hour, we have

$$P_0 = \left[\sum_{n=0}^{1} \frac{1}{n!}\left(\frac{21}{12}\right)^n + \frac{1}{2!}\left(\frac{21}{12}\right)^2 \frac{2.(120/7)}{2(120/7)-30}\right]^{-1}$$

$$= + \frac{7}{4} + \frac{49}{4} = \frac{1}{15}$$

Average waiting time of arrivals in the queue,

$$W_q = \frac{1}{(s-1)!}\left(\frac{\lambda}{\mu}\right)^s \frac{\mu}{(s\mu-\lambda)^2}.P_0 = \left(\frac{7}{4}\right)^2 \frac{120/7}{(120/7-30)^2} \times \frac{1}{15}$$

$$= \frac{343}{30} \times 60 \text{ hour or } 11.43 \text{ minutes.}$$

Example 28(a):

Trains arrived the yard every 15 minutes and the service time is 33 minutes. It the line capacity of the yard is limited to 4 trains, find

(i) *The probability that the yard is empty,*

(ii) *The average number of trains in the system.*

Solution:

λ = 1/15 per minute,

μ = 1/33 per minute.

∴ $\rho = \lambda/\mu = 33/15 = 2.2$,

N = 4.

(i) $p_0 = \frac{1-\rho}{1-\rho^{N+1}} = \frac{\rho-1}{1-\rho^{N+1}} = \frac{2.2-1}{2.2^5-1} = \frac{1.2}{51.5-1} = 0.0237.$

(ii) Average number of trains in the system,

$$L_s = \sum_{n=0}^{N} np_n = 0 + p_1 + 2p_2 + 3p_3 + 4p_4$$

$$= p_0 (\rho + 2\rho^2 + 3\rho^3 + 4\rho^4)$$

$$= 0.0237 [2.2 + 2 \times 2.22 + 3 \times 2.23 + 4 \times 2.24]$$

$$= 0.0237 [2.2 + 9.68 + 31.94 + 83.70] = 3.26.$$

Example 28(b):

In a factory cafeteria the customers (employees) have to through three counters. The customers buy coupons at the first counter, select and collect the snacks at the second counter and collect tea at the third. The server at each counter take son an average 1.5 minutes although the distribution of service time is approximately Poisson at an average rate of 6 per hour. Calculate:

(a) *The average time a customer spends waiting in the cafeteria.*

(b) *The average time of getting the service.*

(c) *The most probable time in getting the service.*

Solution:

From the data of the problem, we have

$\lambda = 6$ customers/hour; Service time per phase = 15 minutes

$\mu = 4.5$ (= 1.5 × 3) customer/minute or 13.34/hour; k = 3

(a) Average time a customer spends waiting in the cafeteria

$$W_q = \frac{k+1}{2k}\,\frac{\lambda}{\mu(\mu-\lambda)} = \frac{3+1}{2(3)}\,\frac{6}{13.34(13.34-6)}$$

$$= \frac{9}{220} \text{ hour or 2.43 minutes}$$

(b) Average time of getting the service is the average of the time when it is following the third phase of service. Thus, average time spent in getting the service is

$$\frac{1}{\mu} = \frac{1}{13.34} \text{ hour or 4.50 minutes}$$

(c) The most probable time spent in getting the service is the model value of service time for the third phase of service. Thus, most probable time spent is

$$\frac{k-1}{k\mu} = \frac{3-1}{3\times 13.34} = \frac{1}{20} \text{ hour or 3 minutes}$$

Example 28(c):

An airline maintenance base has facilities for overhauling only one aeroplane engine at a time. Hence, to return the aeroplanes into use at the earliest, the policy is to stagger the overhauling of the 4 engines of each aeroplane. In other words, only one engine is overhauled each time an aeroplane comes in to the base.

Under this policy, aeroplanes have arrivals according to a Poisson process at a mean rate of one per day. The time required for an engine overhaul has an exponential distribution with mean of half day.

A proposal has been made to change the policy so as to overhaul all four engines consecutively each time an aeroplane comes into the shop. It is pointed out that although this will quadruple the expected service time, each plane would need to come into the shop only one-fourth time as often. Compare the two alternatives on a meaningful basis.

Solution:

The two alternatives will be compared on the basis of the waiting time cost of the aeroplanes requiring overhauling

First alternative {(M/M/1) : (∞/FCFS)} queuing system

Give that λ = 1 aeroplane per day; μ = 2 aeroplanes per day

Therefore, average number of aeroplanes in the system is

$$L_s = \frac{\lambda}{\mu - \lambda} = \frac{1}{2-1} = 1$$

Second alternative {(M/E_k/1) : (∞/FCFS)) queuing system

Give that λ = 1/4 aeroplane per day; k = 4

Since service time per aeroplane is 4 × (1/2) = 2 days, therefore, mean service rate, μ = 1/2 aeroplane per day.

Thus average number of aeroplanes in the system are

$$L_s = \frac{k+1}{2k}\frac{\lambda^2}{\mu(\mu-\lambda)} + \frac{\lambda}{\mu}$$

$$= \frac{4+1}{2(4)} \cdot \frac{(1/4)^2}{1/2(1/2-1/4)} + \frac{1/4}{1/2} = \frac{13}{16} \text{ or } 0.81$$

Since L_s (= 0.81) in the second alternative is less than its value in the first alternative, therefore the waiting cost for requiring overhauling in the second alternative will be less. Hence, the proposal should be accepted.

Example 28(d):

At a certain airport it takes exactly 5 minutes to land an aeroplane, once it is given the signal to land. Although incoming planes have scheduled arrival times the wide variability in arrival times produces an effect which makes the incoming planes appear to arrive in a Poisson fashion at an average rate of 6 per hour. This produces occasional stock up sat the airport which can be dangerous and costly. Under these circumstances, how much time will a pilot expect to spend circling the field waiting to land?

Solution:

From the data of the problem, we have

λ = 6 per hour or 1/10 per minute; μ = 1/5 per minute

k = ∞, as service time is constant

Hence, the average time which a pilot expects to spend circling the field waiting to land is given by

$$W_q = \lim_{k \to \infty} \frac{k+1}{2k} \frac{\lambda}{\mu(\mu-\lambda)}$$

$$= \lim_{k \to \infty} \frac{1}{2}\left(1+\frac{1}{k}\right)\frac{\lambda}{\mu(\mu-\lambda)}$$

$$= \frac{1}{2}\left(1+\frac{1}{\infty}\right)\frac{1/10}{1/5(1/5-1/10)}$$

= 5/2 or 2.5 minutes.

Example 29:

A mechanic repairs four machines. The mean time between service requirements is 5 hours for each machine and forms an exponential distribution. The mean repair time is hour and also follows these me distribution pattern. Machines downtime cost Rs. 25 per hour and the mechanic costs Rs. 55 per day.

(a) *Find the expected number of operating machines.*

(b) *Determine the expected downtime cost per day.*

(c) *Would it be economical to engage two mechanics, each repairing only two machines?*

Solution:

This situation involves finite population.

Arrival rate λ = 1/5 = 0.2,

service rate μ = 1/1 = 1.

Let us first find the probability of an empty system,

$$p_0 = \frac{1}{\sum_{n=0}^{n=4} \frac{4!}{(4-n)!}\left(\frac{2}{1}\right)^n}$$

$$= \frac{1}{1+4(.2)+(4\times3)(.2)^2+(4\times3\times2)\times(.2)^3+(4\times3\times2\times1)(.2)^4} = 0.4.$$

(a) Expected number of broken-down machines in the system,

$$Ls = M - \frac{\mu}{\lambda}(1-p_0)$$

$$= 4 - \frac{1}{2}(1-4) = 4 - 5 \times .6 = 4 - 3 = 1.$$

$\therefore$ Expected number of operating machines in the system = 4 – 1 = 3.

(b) Expected down time cost per day (assuming an 8-hour day)

= 8 × expected number of broken-down machines × Rs. 25 per hour.

= 8 × 1 × 15 = Rs. 200 per day.

(c) When there are two mechanics each serving two machines, M = 2.

$$\therefore \quad p_0 = \frac{1}{\sum_{n=0}^{n=2} \frac{2!}{(2-n)!}\left(\frac{2}{1}\right)^n} = \frac{1}{1+2(.2)+2\times1(.2)^2} = \frac{1}{1.48} = 0.68.$$

It is assumed that each mechanic with his two machines constitutes a separate system with no interplay. Expected number of machines in the system

$$= M - \frac{\mu}{\lambda}.(1-p_0) = 2 - \frac{1}{2}(1-.68) = 0.4.$$

$\therefore$ Expected down time/day = 8 × .4 × number of mechanics

= 8 × .4 × 2 = 6.4 hr/day.

$\therefore$ Total cost with two mechanics

= Rs. 2 × 55 + Rs. 6.4 × 25

= Rs. (110 + 160) = Rs. 270 per day.

Total cost with one mechanic = Rs. (55 + 200) = Rs. 255 per day.

Hence use of two mechanics is not economical.

Example 30:

Arrivals at telephone booth are considered to be Poisson with an average time of 10 minutes between one arrival and the next. The length of phone call is assumed to be distributed exponentially, with mean 3 minutes.

(a) *What is the probability that a person arriving at the booth will have to wait?*

(b) *The telephone department will install a second booth when convinced that an arrival would expect waiting for a least 3 minutes for a phone call. By how much should the flow of arrivals increase in order to justify a second booth?*

(c) *What is the average length of the queue that forms from time to time?*

(d) *What is the probability that it will take him more than 10 minutes altogether to wait for the phone and complete his call?*

Solution:

From the data of the problem, we have

$\lambda = 1/10 = 0.10$ person per minute and $\mu = 1/3 = 0.33$ person minute

(a) Probability that a person has to wait at the booth.

$$P(n > 0) = 1 - P_0 = \lambda/\mu$$
$$= 0.10/0.33 = 0.3.$$

(b) The installation of second booth will be justified only if the arrival rate is more than the waiting time. Let l' be the increased arrival rate. Then expected waiting time in the queue will be

$$W_q = \frac{\lambda'}{\mu(\mu - \lambda')}$$

$$3 = \frac{\lambda'}{0.33(0.33 - \lambda')} \text{ or } \lambda' = 0.16$$

where $W_q = 3$ (given) and $\lambda = \lambda'$ (say) for second booth.

Hence, the increase in the arrival rate is 0.16 – 0.10 = 0.06 arrivals per minute.

(c) Average length of non-empty queue

$$L_b = \frac{\mu}{\mu - \lambda} = \frac{0.33}{0.23} = 2 \text{ customers (approx).}$$

(d) Probability of waiting for 10 minutes or more is given by

$$P(t \geq 10) = \int_{10}^{\infty} \frac{\lambda}{\mu}(\mu - \lambda)e^{-(\mu-\lambda)t}\, dt$$

$$= \int_{10}^{\infty} (0.3)(0.23)e^{-0.23t}\, dt = 0.069 \left[\frac{e^{-0.23t}}{-0.23}\right]_{10}^{\infty} = 0.03$$

This shows that 3 per cent of the arrivals on an average will have to wait for 10 minutes or more before they can use the phone.

Example 31:

A warehouse has only one loading dock manned by a three person crew. Trucks arrive at the loading dock at an average rate of 4 trucks per hour and the arrival rate is Poisson distributed. The loading of a truck takes 10 minutes on an average and can be assumed to be exponentially distributed. The operating cost of a truck is Rs. 20 per hour and the members of the loading crew are paid @ Rs. 6 each per hour. Would you advise the truck owner to add another crew of three persons?

Solution:

From the data of the problem, we have $\lambda = 4$ per hour; $\mu = 6$ per hour

Total hourly cost = Loading crew cost + Cost of waiting time.

= {(Number of loaders) × (Hourly wage rate)}

+ {(Expected waiting time per truck, Ws) (Expected arrival per hour, λ)

$$= 6 \times 3 + \frac{1}{6-4} \times 4 \times 20 = \text{Rs. } 58 \text{ per hour.}$$

$$\text{Total hourly cost} = 6 \times 6 + \frac{1}{12-4} \times 4 \times 20 = \text{Rs. } 46 \text{ per hour.}$$

Since the total hourly cost after addition of another crew of three persons is less than the existing cost, therefore the truck owner must add a crew of another 3 loaders.

EXERCISES

1. Customers arrive at a box office window, being manned by a single individual according to a Poisson input process with a mean rate of 30 per hour. The time required to serve a customer has an exponential distribution with a mean of 90 seconds. Find the average waiting time of a customer. Also determine the average number of customers in the system and average queue length.
2. The mean rate of arrival of planes at an airport during the peak period is 20 per hour, and the actual number of arrivals in any hour follows a Poisson distribution. The airport can land 60 planes per hour on an average in good weather and 30 planes per hour in bad weather, but the actual number landing in any hour follows a Poisson distribution with these respective averages. When there is congestion, the planes are forced to fly over the field in the stack awaiting the landing of other planes that arrived earlier.
   (a) How many planes would be flying over the field in the stack on an average in good weather and in bad weather?
   (b) How long would a plane be in the stack and in the process of landing in good and in bad weather.
3. A repair shop attended by a single mechanic has an average of four customers an hour who bring small appliances for repair. The mechanic inspects them for defects and takes six minutes on an average. Arrivals are Poisson and service rate has the exponential distribution. You are required to

(a) Find the proportion of time during which there is no customer in the shop.

(b) Find the probability of finding at least one customer in the shop.

(c) What is the average number of customers in the system?

(d) Find the average time spent by a customer in the shop including service.

4. In a bank cheques are cashed at a single 'teller' counter. Customers arrive at the counter in a Poisson manner at an average rate of 30 customers per hour. The teller takes, on an average, a minute and a half to cash cheque. The service time has been shown to be exponentially distributed:

   (a) Calculate the percentage of time the teller is busy.

   (b) Calculate the average time a customer is expected to wait.

5. In a tool crib manned by a single assistant, operators arrive at the tool crib at the rate of 10 per hour. Each operator needs 3 minutes on an average to be served. Find out the loss of production due to the time lost in waiting for an operator in a shift of 8 hours if the rate of production is 100 units per shift.

6. Discuss the fields of application for queuing theory. Explain queue discipline and its various forms.

7. Explain the basic queuing process. What are the important random variates in queuing system to be investigated?

8. What is queuing theory? What types of questions are sought to be answered in analysing a queuing system?

9. What do you understand by a queue? Give some important applications of queuing theory?

10. Telephone users arrive at a booth following a Poisson distribution with an average time of 5 minutes between one arrival and the next. The time taken for a telephone call is on an average 3 minutes and it follows an exponential distribution. What is the probability that the booth is busy? How many more booths should be established to reduce the waiting time to less than or equal to half of the present waiting time.

11. At a railway station, only one train is handled at a time. The railway yard is sufficient only for two trains to wait while the other is given signal to leave the station. Trains arrive at the station at an average rate of 6 per hour and the railway station can handle them on an average of 12 per hour. Assuming Poisson arrivals and exponential

service distribution, find the steady-state probabilities for the various number of trains in the system. Also find the average waiting time of a new train coming into the yard.

12. Patients arrive at a clinic according to a Poisson distribution at the rate of 30 patients per hour. The waiting room does not accommodate more than 14 patients. Examination time per patient is exponential with mean rate of 20 per hour.
    (i) Find the effective arrival rate at the clinic.
    (ii) What is the probability that an arriving patient will not wait? Will he find a vacant seat in the room?
    (iii) What is the expected waiting time until a patient is discharged from the clinic?
13. Queuing theory can be used effectively in determining optimal service levels. Elucidate this statement with the help of an example.
14. What is a queuing theory problem? Describe the advantages of queuing theory to a business executive with a view to persuading him to make use of the same in management.
15. Assume that goods trains are coming in a yard at the rate of 30 trains per day and suppose that the inter-arrival times follow an exponential distribution. The service time for each train is assumed to be exponential with an average of 36 minutes. If the yard can admit 9 trains at a time (there being 10 lines one of which is reserved for shunting purpose). Calculate the probability that the yard is empty and find the average queue length.